The Advisors

OPPENHEIMER, TELLER, AND THE SUPERBOMB

Herbert F. York
UNIVERSITY OF CALIFORNIA, SAN DIEGO

Contents

Preface

Germany invaded Poland and started World War II just two weeks before I entered the University of Rochester in September 1939. Ever since then, my professional life has been dominated by the nuclear arms race.

During my sophomore and junior years, my physics professors, including Lee A. DuBridge and Victor Weisskopf, began to disappear one by one into secret war laboratories. After my own accelerated graduation in May 1943, I too left for the University of California Radiation Laboratory in Berkeley, where we were engaged in the separation of uranium isotopes. The director of the laboratory was Ernest O. Lawrence, and my immediate boss was Frank Oppenheimer. Frank's older brother, J. Robert Oppenheimer, was at that time establishing a laboratory at Los Alamos, New Mexico, where the isotopes we were separating would, two years later, be fabricated into history's first atomic weapon.

For the next quarter century, with only a very brief respite following the end of World War II, I participated directly and centrally in the nuclear arms race.

In 1951 I was a participant in Operation Greenhouse at Eniwetok when the very first thermonuclear experiments were conducted, and I was subsequently (1952–1958) the first director of the Lawrence Livermore Laboratory where much of the work to exploit those early results was carried out. Later, just after Sputnik and during the height of the "missile gap" crisis, I was briefly the first chief scientist of the Advanced Research Projects Agency (1958) and then for a longer period the first director of Defense Research and Engineering (1958–1961).

In addition to these full-time activities, I also have served on many of the advisory committees that helped mold American plans and policies during the most dynamic phases of the arms race. Among the most crucial were the several "Von Neumann" committees (1953–1956), which helped to mold the mainstream U.S. missile and rocket programs, the President's Science Advisory Committee (1957–1958 and 1964–1968), and the General Advisory Committee on Arms Control and Disarmament (1962–1969). Richard Nixon's acceptance of my formal resignation from this last committee in April 1969 ended my twelve years of direct service to the executive branch of the U.S. government. Since then I have continued to testify occasionally before congressional committees on matters relating to the arms race, and I have actively participated in a number of closely connected nongovernmental organizations designed to influence the course of events at the interface between technology and public affairs, including the Federation of American Scientists and the international Pugwash movement.

In 1969, I wrote a book which summarized my views about the course of the arms race during the prior quarter century and my judgments about what is portended for the future. It is called *Race to Oblivion: A Participant's View of the Arms Race* (Simon and Schuster, New York, 1970). In the last chapter I concluded;

> Our unilateral decisions have set the rate and scale for most of the individual steps in the strategic-arms race. In many cases we started development before they did and we easily established a large and long-lasting lead in terms of deployed numbers and types. Examples include the A-bomb itself, intercontinental bombers, submarine-launched ballistic missiles, and MIRV. In other instances, the first development steps were taken by the two sides at about the same time, but immediately afterward our program ran well

> ahead of theirs both in the development of further types and applications in the deployment of large numbers. Such cases include the mighty H-bomb and, very probably, military space applications. In some cases, to be sure they started development work ahead of us and arrived first at the stage where they were able to commence deployment. But we usually reacted so strongly that our deployments and capabilities soon ran far ahead of theirs and we, in effect, even here, determined the final size of the operation. Such cases include the intercontinental ballistic missile and, though it is not strictly a military matter, manned space flight.

In making such assertions I do not charge that the United States bears the major part of the blame for the existence of either the cold war or the technological arms race; in my view the responsibility for these is widely shared among the major powers of the world. However, I do believe that the United States has pursued policies which caused the technological arms race to advance at a substantially faster pace than was really necessary for America's own national security. The reasons for this are not that American leaders have been less sensitive to the dangers of the arms race than the leaders of other countries, nor that they are less wise or more aggressive. Rather, the reason is that the United States is richer and more powerful, and its science and technology are more dynamic and generate more ideas and inventions of all kinds, including ever more powerful and exotic means of mass destruction. In short, the root of the problem has not been maliciousness, but rather a sort of technological exuberance that has overwhelmed the other factors that go into the making of overall national policy.

This book is about one of those particular instances: the development of the H-bomb, or the superbomb as it was then called. It is an especially important instance because it is one of the relatively few cases where those who explicitly tried to moderate the nuclear arms race came within shouting distance of doing so.

It seems clear that if humanity is to survive, those who seek first to slow and stop the arms race and then to reverse it must succeed before there is too much more technological "progress" of the kind described in this book. Understanding earlier cases where the voices of moderation almost succeeded may contribute to hastening the day when they really will.

During the preparation of this work I have talked and corresponded with many of the people mentioned herein, including Luis W. Alvarez, Hans Bethe, Norris E. Bradbury, Harold Brown, Lee A. DuBridge, J. Carson Mark, K. D. Nichols, Edward Teller, and Stanislaw Ulam. The Atomic Energy Commission's official historian, Richard G. Hewlett, has been of great help, and so have three graduate students in history here at the University of California, San Diego: Daniel Melcher, Gregory Grebb, and Renee Leap. I thank all of them for their cooperation, but, of course, I am solely responsible for what follows.

May 1975 *Herbert F. York*

The Advisors

Chapter ONE

Introduction

In 1948 Czechslovak Communists carried out a coup in the shadow of the Red Army and replaced the previous government of that country with one subservient to Moscow. Also in 1948, the Russians unsuccessfully attempted to force the Western allies out of Berlin by blockading all land transport routes to the city. In early 1949 Mao Tse-tung's People's Liberation Army captured Peking and soon after established the People's Republic of China. Taken together these and similar but less dramatic events were generally perceived in the West as resulting in the creation of a monolithic and aggressive alliance stretching the full length of the Eurasian continent, encompassing almost half the world's people, and threatening much of the rest. Then, in the fall of 1949, the Soviets exploded their first atomic bomb and ended forever the brief American nuclear monopoly. Coming as it did at a time when virtually all Americans saw the cold war as rapidly moving from bad to worse, this event was seen as a challenge that demanded a reply. The immediate challenge being nuclear, an especially intensive search for an appropriate response was conducted by American nuclear circles.

Most proposed responses involved substantial but evolutionary changes in current American programs: expand the search for additional supplies of ore, step up the production of weapons, adapt nuclear weapons to a broader range of delivery vehicles and end uses, and the like. One proposal was radically different. It called for the fastest possible development of the so-called superbomb, or hydrogen bomb. This bomb was based on an entirely new and as yet untested principle, and was estimated to be one thousand or more times as powerful as the atomic bombs that had marked the end of World War II. Work on the theory of the superbomb had already been going on for seven years, but it had never had a very high priority, and so far it had yielded no practical result. A number of scientists and politicians endorsed the proposal, but for years Edward Teller had been and still remained its leading advocate.

The General Advisory Committee (GAC) of the Atomic Energy Commission (AEC) reviewed all of these proposals. The committee, which was chaired by Robert Oppenheimer, agreed that the United States ought to accelerate both the production and development of atomic bombs generally, but concluded that it should forgo the development of the superbomb.

A brief, intense, highly secret debate resulted.

The opponents of the superbomb argued that neither the possession of a super weapon nor the initiation of its development was necessary for maintaining national security and that under such circumstances it would be morally wrong to initiate the development of such an enormously powerful and destructive weapon. In essence, they argued that the world ought to avoid the development and stockpiling of the superbomb if at all possible, and that America's forgoing it was a necessary precondition for persuading others to do likewise. Furthermore, they concluded that the relative status and dynamism of American nuclear technology were such that the United States could safely run the risk that the Soviet Union might not practice similar restraint and would instead initiate a secret program of its own.

The advocates of the superbomb contended that the successful achievement of such a bomb by the Soviets was only a matter of time, and so at best our forgoing it amounted to a deliberate decision to become a second-class power and at worst it was tantamount to surrender to the dark forces of world communism. They added that undertaking its development was morally no different from developing any other weapon.

Even though the two main protagonists, Robert Oppenheimer and Edward Teller, knew each other well, the debate did not take the form of a direct confrontation between them. Instead, each in his own way and with his own group of allies sought to persuade higher authorities to his particular point of view. Oppenheimer did so by very effectively using the several formal and direct channels provided by his membership on major official advisory committees. Teller had no such formal channels, but he did have several good friends and admirers in high places who arranged for him to present his ideas and opinions to higher authorities on a purely personal basis. Both men possessed brilliant but qualitatively different intellects, and both had charismatic personalities. Oppenheimer was able quickly to analyze and understand complex situations, and to synthesize a wide spectrum of separate ideas and concepts into an overarching scheme. Teller, a brilliant inventor of new ideas, had a special way of tirelessly and relentlessly turning over in his mind questions that, for whatever reason, bothered or intrigued him.

In several important ways, the two men and their backgrounds were much alike. Both were theoretical physicists, and within physics their particular fields of interest overlapped. Both had cultural and political interests that went well beyond the confines of their principle vocations. They were of the same generation (Oppenheimer was less than four years older than Teller) and both were born into affluent and cultured Jewish families. Both made exceptionally strong impressions on people they came into close contact with, from students to presidents, and these impressions could be either positive or negative.

In other important respects, they were different. Oppenheimer was born and raised in America. Originally apolitical, the anti-Jewish horrors of the Hitler regime awakened his social conscience, and the militant opposition of the Communists to nazism and fascism, which became particularly evident during the Spanish civil war, brought him into close contact with various American left-wing groups. Although never a member of the Communist party himself, many of his friends, some of his students, his wife, and his younger brother were at one time or another. As a result of his great store of knowledge and his high intelligence, he could be extremely persuasive in a discussion or an argument, whether in a seminar room or at a congressional hearing, but he was often arrogant and he could make cutting and even cruel remarks about those who disagreed with him and their views. Over the years, the list of people who had felt the almost physical force of his disdain came to include a number of people in high and influential positions. The most important of them for the purposes of this history was Lewis L. Strauss, a member of the AEC during the debate over the superbomb and its chairman at the time of Oppenheimer's final downfall.

Teller was born in Hungary shortly before World War I broke out, and he was raised there in circumstances that caused him to be acutely aware of the Russian army as a military threat (Hungary and Russia were enemies with a long common border during World War I) and the Communist party as a political threat (there was a briefly successful Communist regime in Hungary in 1919). He came to America in 1935, part of the intellectual migration that resulted from the spread of Hitlerism across the face of central Europe. Some of his relatives remained in Hungary, however, and years later, while the events described in this book were going on, he often expressed concern over their situation under thé new Communist regime that had just recently been reestablished there in the shadow of Soviet military power. Like Oppenheimer, his knowledge and wit made him extremely persuasive, and his arguments were also sometimes tinged with arrogance, but he never to my knowledge made cutting or cruel comments. Moreover, in any serious discussion or debate, whether in a classroom, a congressional hearing, or simply face-to-face with one other individual, Teller usually exuded a sort of boyish enthusiasm which, coupled with a typically central European charm and even diffident manner, favorably impressed most people, especially politicians and statesmen, and predisposed them to believe what he was telling them. Over the years since he began dealing in the political arena (that is, since shortly after the end of World War II) he has built up a substantial list of warm admirers in high places, some of whom regard him with an almost religious awe.

In January 1950 President Harry S. Truman resolved the debate in favor of the proponents of the superbomb. Two years and nine months later the United States exploded the first superbomb, code-named Mike, at Eniwetok Atoll. It yielded ten megatons of explosive energy, which was the amount that had long been predicted, just about a thousand times as much as the bomb which had killed a hundred thousand people at Hiroshima seven years before. Three years and a few weeks after Mike, the Soviets exploded a somewhat smaller superbomb in central Asia.

The rapid and successful development of the superbomb (or super as it came to be called) was not the only way in which the arms race was accelerated by the complex web of events and decisions that started from the first Soviet atomic explosion and ended with Truman's decision. In addition, this web of interacting events resulted in a large proliferation of both numbers and types of ordinary nuclear weapons, in the establishment of a second U.S. nuclear weapons laboratory in 1952, and in the removal of Oppenheimer's security clearance in 1954 on the grounds that he was a security risk.

The side in the weaker formal position won the debate. Largely as a result of this peculiar situation, soon after Truman's decision some of Oppenheimer's enemies initiated attempts to remove him from all positions of influence over nuclear affairs. These attempts were ultimately successful. In late 1953, he was officially charged with being a security risk and his remaining clearances were summarily canceled. Oppenheimer denied the charge, and in the spring of 1954 a special security hearing was held to decide the matter. The final result was that the charge was upheld. This hearing in particular has attracted a great deal of attention in the intervening years, and there are many excellent studies of it as an object lesson in political morality and as an example of the effect of grossly exaggerated notions of the role of secrecy in maintaining national security. In my view one of the best of these studies is Philip Stern's *The Oppenheimer Case.*

This book has quite a different purpose. It examines in some detail the intellectual content of the debate itself. It places particular emphasis on the technical and strategic elements of the GAC's advice, and judges its soundness both by comparing its technological predictions with what actually occurred and also by estimating what would have happened if Truman had followed it and contrasting that with what did actually happen. In doing so, I confine myself largely to the question of how the nuclear balance would have been affected over the course of the next half decade or so. In that restricted case, the technological facts that constrain the various possible alternative worlds are relatively few in number and by now well enough known so as to make it possible to determine in retrospect what the most favorable, the most likely, and the worst plausible alternative worlds actually were. To summarize what follows, my conclusions are that the GAC's advice was basically sound, its predictions were remarkably close to the mark, and the nuclear balance would not have been in serious danger of being upset had we followed it. The 1954 persecution of Oppenheimer and the official finding that he was a security risk cannot, therefore, be *ex post facto* justified even on the purely practical grounds that his advice was so dangerous and his influence so great that his complete removal from the advisory system was necessary for national security reasons.

On the other hand, in this book I do *not* attempt to work out how the overall course of the cold war might have changed had Truman followed the GAC's advice. There the alternatives are far too many and too diverse for me to attempt to assess them. Nor do I attempt to bring out all of the manifold influences acting on Truman at the time he made his decisions. There were many matters besides those which were brought out by either Oppenheimer and his allies or Teller and his, and I cannot, therefore, fully assess on the basis of the material presented in this book what Truman might have perceived his real options to be. I can however, say with certainty that when Truman rejected the GAC's advice, we all missed (as the GAC's report put it) one of the rare "opportunit[ies] of providing by example some limitation in the totality of war and thus limiting the fear and arousing the hopes of mankind."

In all of this, I have not gone deeply into the underlying personalities or motives, but I have tried to provide a necessary minimum of background information about who the principals were, and I have briefly mentioned what I knew (and did not know) at the time and what was going on in my own mind at a few key junctures. I have concentrated on the arguments and the events and I have tried to analyze their interrelationships in sufficient depth to bring out the causal relationships among them. In the course of doing so, I have not only described both the American and Soviet superbomb programs, I have also followed up some of the lesser known, but still important, consequences that flowed from the American determination to respond adequately to the first Soviet bomb, including the proliferation of weapons and the creation of the second laboratory.

I knew personally nearly all of the principals in these events, though not at all equally well. I had a close teacher–student relationship with Lawrence that stretched over fifteen years, I worked as a colleague of Edward Teller for eight years, I took graduate physics courses from Robert Oppenheimer, and I met frequently with Norris Bradbury and Lewis Strauss. I participated in some of the most important and most relevant events, and I watched some of the others from a front-row seat.

I had joined Lawrence's laboratory in 1943 in its work on the preparation of the nuclear explosives used in the Hiroshima bomb. After the war, I remained at Berkeley as a graduate student, and gained my Ph.D. in 1949 for work I did using the facilities and equipment of the same laboratory. When the first Soviet bomb exploded, I was just beginning a career in basic research. At the time the debate over the superbomb was actually going on, I was almost completely unaware of it, but very soon after it was resolved I was, as will be described below, gradually drawn into the center of the superbomb program and many of the other programs and projects that came out of the same nexus of events.

Even so, this book is not in any important sense a memoir. I have used my familiarity with the people and my direct knowledge of most events only as a starting point for the preparation of this history. By far the largest part of the factual information in what follows is drawn from or based on documentary material, and most of the rest comes from recent conversations with several of the surviving principals. Only a few items, none of them essential to the story, are based solely on my own memory of events.

Much of the most important documentary material has become available only very recently. Foremost among this material is the full report of the GAC's 1949 deliberation on the superbomb. Except for some very brief extracts which appeared in the records of the 1954 Oppenheimer hearing, the report only became declassified in 1974, and even then a few strictly technical data were still omitted. The full report, less these minor deletions, is presented in the Appendix of this book. Other recent documents essential to this study are two biographies of Igor V. Kurchatov (the scientific director of the Soviet program), one by P. Astashenkov and one by Igor N. Golovin (a scientific colleague of Kurchatov's), published in Moscow in 1967 and 1969, respectively; a very heavily censored version of the report of the 1946 conference on the super, released in 1971; a brief technical history of the early thermonuclear program released in 1974, but written in its original version in 1954 by Carson Mark with full access to the original research documents; and Stanislaw Ulam's *Adventures of a Mathematician* (1976). in which he gives his recollections about these events. Older essential documents are Teller's *The Legacy of Hiroshima;* the official transcripts of the Oppenheimer hearing; the two official AEC histories, *The New World* by Hewlett and Anderson, and *A History of the U.S. Atomic Energy Commission* by Hewlett and Duncan; and some papers in the E. O. Lawrence archives released in the 1960's but unnoticed at the time

There is, I think, only one truly central technological fact in all this that still remains secret, and that is the precise nature of the Teller–Ulam invention of 1951. From the point of view of understanding what happened, it is less than ideal to omit it, but I agree with the security officials of the five thermonuclear powers (U.S.A., U.S.S.R., U.K., France, and China) that this particular detail should remain as restricted as possible in its circulation. Moreover, I believe that even without it, a sufficiently complete and coherent story can be put together to make it possible to understand both what did actually happen and what might have been.

The history that emerges from a careful examination of the newly available information is in a number of important aspects quite different from that which has previously made its way into American folklore.

The following is a summary of some of the new elements that are explained more fully in the following chapters. Even those who are quite familiar with the prior available literature on the subject will find some important surprises here.

- At the time the GAC wrote its October 1949 report, there were not one but two importantly different thermonuclear programs under way at the Los Alamos laboratory. One of these was the so-called booster program, which had as its main practical objective improving the efficiency of fission bombs. The other was the "super" program, which had as its objective producing a weapon having a power at least a thousand times greater than that of the then standard nuclear bomb. The full report makes it clear that the GAC strongly endorsed the former as being potentially very useful and only opposed the latter as being murderous to a degree far beyond that which could be justified even within the intellectual framework that took a large stock of nuclear weapons as being absolutely essential for national security. Thus, the GAC did not, as is sometimes charged, try to stop "the progress of science" or the acquisition of new knowledge about thermonuclear processes. They only tried to stop, or at least delay, what they took to be a particularly horrendous and uncalled-for application of such new knowledge.
- The full report shows that the GAC was realistic about what might flow from a decision on our part to forgo the development of the super. At no point do its authors imply that following their advice would prevent the development of the super by someone else. Rather, they clearly recognize the simple fact that our forgoing the super was a necessary precondition for persuading others to do so and they correctly estimated that our then current and planned nuclear capability was such that our doing so would not entail any substantial risk of upsetting the current balance of nuclear power. Far from being naive, the minority addendum to the report makes it clear that the committee considered the possibility that the Russians might even go so far as to make an agreement on the super and then renege on it secretly.
- The GAC report did not at all espouse what would be called in now current terms a "dovish" line. It is true that most of the members in varying degrees did favor some effective form of

nuclear arms control; and two of the members, Robert Oppenheimer and I. I. Rabi, were earlier the authors and promoters of the particular set of ideas that eventually came to constitute the heart of the Baruch plan. However, in the absence of any effective agreements they also strongly favored maintaining and improving sufficient nuclear power to guarantee fully our national security. They therefore endorsed a series of proposed AEC actions which were designed to produce more and better nuclear weapons and to prepare for the use of these weapons in a broader set of military situations. In retrospect these actions were certainly not less, and were probably substantially more than the actual situation as we now know it called for. As Gilpin[1] later put it, the internal debate over the super was between "the finite containment school" and the "infinite containment school." Or, in today's terms the argument was between moderates and hawks, or perhaps between hawks and superhawks. No full-fledged doves, or what Gilpin calls members of the "Arms Control School" were involved for the simple reason that none of them had the necessary clearances.

▪ In the past, much significance has been attached to the first Soviet thermonuclear explosion in August 1953. The Soviets have called it the "first-in-the-world hydrogen bomb" and most Western political, historical, and technical writers have accepted their statement at face value and have assumed that the Soviets in some important sense "won" the "race" for super. The discussions of the U.S. and the Soviet program presented in this book make it clear that this was not the case and that the United States was at all times well ahead of the Soviet Union in the development of the super. In particular, the August 1953 Soviet explosion was of a device which was no more powerful than the largest prior U.S. fission explosion and only something like three to five percent as powerful as the first American thermonuclear explosion, both of which had taken place nearly a year earlier. Moreover, the August 1953, Soviet device was not based on design principles (like the Teller–Ulam invention of 1951) which lead directly to the construction of bombs of indefinitely large size. In short it was not a super; the first Soviet super-like device came only in late 1955, three years after the United States had first experimentally verified the basic ideas that made the very large hydrogen bomb practical. Since the U.S. and the Soviet program followed different paths to the same ultimate goal, it is not possible to say exactly in years how far ahead the United States was, but the lead was always very substantial, and a reasonable understanding of the nature of this lead can be derived from the details presented below in chapters 5 and 6 (and hence, without knowing exactly what the Teller–Ulam invention is).

■ The first Soviet A-bomb and the U.S. determination to react to it led by a somewhat complex path to the creation of a second American nuclear weapons laboratory at Livermore, California, and hence eventually to a doubling of the size of the American nuclear weapons development program. Some of the elements of the story of how this happened have been presented before, but the history given here, while brief, is more comprehensive. This particular episode, like the history of the super itself, can be seen as an illustration of just how what Secretary of Defense McNamara called technological momentum can determine the course of the arms race. The possibilities that welled up out of the technological program and the ideas and proposals put forth by the technologists eventually created a set of options that was so narrow in the scope of its alternatives and so strong in its thrust that the political decision makers had no real independent choice in the matter.

■ In the course of presenting its case, the GAC made a number of predictions about some of the properties of the super, and about the path which a program to develop it would follow. While not perfect, they are remarkably accurate considering the difficulties inherent in making them.

■ The main body of the GAC report makes a plea for lowering as much as possible the barriers of secrecy that surrounded the whole process of deciding what to do about the super. This same plea was echoed over and over in statements made by other scientists in this same context, and Oppenheimer returned to it often in subsequent articles and speeches. The committee recognized that certain technical details should be kept "secret," but they felt a very large part of what they were discussing could be made available without endangering the national security. They obviously felt very strongly that such momentous decisions affecting all mankind should not be made by a tiny elite in-group exclusively privy to all of the relevant facts, even though in this case they were themselves included in it.

Chapter TWO

The American Nuclear Program to 1949

Origins

Nuclear energy was discovered by Henri Becquerel in France in 1896. He found, largely by accident, that ores containing uranium emitted a form of radiation that could penetrate several layers of opaque paper and fog a photographic emulsion. It was soon realized that enormous amounts of energy were locked up in the nuclei of atoms, amounts of energy exceeding those released in ordinary chemical processes by a millionfold.

For the next forty years, scientists in many countries worked to elucidate the process. In brief, they found that nuclear energy was being very slowly released by the nuclei of ten or so naturally occurring radioactive elements, but that absolutely nothing that they tried had any discernible effect on the rate or on the amount. Then in the 1930's, progress in the field greatly accelerated. In France, Frederic Joliot-Curie discovered artificial radioactivity. In England, James Chadwick discovered the neutron. In Italy, Enrico Fermi used the neutron to explore nuclear processes. And in America, Ernest Lawrence developed the large particle accelerators, or "atom smashers," that enabled nuclear research to advance even more rapidly.

Finally, in December of 1938, Otto Hahn and Fritz Strassmann in Germany discovered the process known as uranium fission. That process can be initiated by a single neutron striking a uranium nucleus. When the process occurs, an enormous amount of energy is released and two or three new neutrons are emitted. Each of these new neutrons can, under just the right conditions, repeat the process, and so an explosive chain reaction can result.

Nuclear physicists all over the world quickly recognized that this discovery could, for the first time, make it possible for man to influence the rate of release of nuclear energy, and to produce either enormous explosions or electric power, depending on just how fast the chain reaction developed.

As a result, in early 1939, leading scientists in Britain, France, Germany, Russia, and the United States focused their attention on the newly discovered process, and began to carry out experiments designed to explore the unprecedented possibilities inherent in it. Then in September 1939, the Nazi armies began to roll, and the whole situation underwent another major change. The rather small French program was first transferred to Britain, and then in due course to Canada. The somewhat larger British program prospered long enough to produce results which convinced the government of the possibilities involved, but then at that point, under the influence of the German aerial bombardment and the press of other more immediate tasks, part of it was transferred to Canada, and the remainder went to the United States where it was melded into the American effort. The Russian program, which never involved more than some tens of people, was completely stopped when the Germans invaded the Soviet Union in the spring of 1941. It was revived on a modest scale about two years later, but it did not come to fruition until long after World War II ended. (The Russian program is described in detail in the next chapter.) The German program never got off the ground. Not only had German science been decimated and thrown into a chaotic state by the expulsion and imprisonment of the Jewish scientists, but the belief that the war would be very short and political factionalism prevented what remained from doing anything much. Only in the United States, then still isolated by the broad Atlantic from the European violence, and with its scientific capacity greatly augmented by the influx of refugees, did the program to exploit the new discoveries have a chance to prosper and produce practical results.

On the basis of much intervening exploratory work in the United States and Britain, on December 6, 1941, the day before the attack on Pearl Harbor, Vannevar Bush, who was Franklin D. Roosevelt's chief scientific aide, signed a report declaring the atomic bomb to be feasible and recommending a crash program for its development. During the first half of 1942, while exploratory nuclear experiments continued at a number of different sites, detailed plans were worked out for organizing the work on the bomb itself, and in the latter half of the year, the Manhattan project was formally instituted and got under way. The main research center for the program was established in early 1943 on an isolated mesa at Los Alamos, New Mexico.

The Los Alamos Scientific Laboratory

The Los Alamos Scientific Laboratory was operated under a contract from the U.S. Army to the University of California. The director was Robert Oppenheimer, previously a professor of theoretical physics at the University's Berkeley campus. Other major nuclear research programs were established at the radiation laboratory in Berkeley under the direction of Ernest Lawrence, at Columbia University under Harold Urey and John Dunning, at the University of Chicago under Arthur Compton, and at the University of Iowa under Frank Spedding. These other institutions worked mainly on questions relating to the production of the basic nuclear explosive materials (U-235 and plutonium), while the Los Alamos laboratory worked on the means for actually making a bomb out of these materials.

The scientific teams recruited for these programs were quite unlike anything ever seen before in terms of both the talents and the numbers of the people making them up. The leaders of American nuclear physics and other relevant sciences, accompanied by their most promising students, flocked to the new research centers. A smaller but singularly talented group of refugee scientists, who had fled from the Hitler terror that had been building up in Europe over the past several years, joined them; and so did a special team from Britain that had been officially sent over to participate in the American program[1].[1] At its peak, the total effort devoted to the project, measured in terms of the rate of dollar expenditures as percent of gross national product, even exceeded slightly the maximum rate experienced during the Apollo Lunar flight program of the 1960's.

The Los Alamos laboratory, which was the largest and most remarkable of all the atomic research centers, in only two years succeeded in developing, designing, and building two different kinds of fission weapons, one using U-235 as its nuclear explosive, and the other using plutonium. The design of the U-235 bomb was based on particularly simple and straightforward principles, and as a result, it was used to destroy Hiroshima on August 6, 1945, without a prior test of whether or not it would work. The plutonium bomb was based on more novel design principles, and it was consequently secretly tested in New Mexico in July before being exploded over Nagasaki on August 9.[2] The two explosions in Japan together killed well over 100,000 people, and played a major part in precipitating the formal surrender of Japan that ended World War II.

When the war ended, most of the scientists on the remote Los Alamos mesa, just like the troops scattered throughout the world, returned "home" to the universities and laboratories from whence

[1] Numbers in brackets refer to source notes in the back of the book.

[2] The design of all varieties of fission bombs is based on the remarkable fact that if a mass of fissile material in excess of the amount known as the "critical mass" (some tens of pounds) is somehow assembled all in one solid piece, it will promptly and automatically explode with great violence. Hence, the basic design simply involves a means for taking an assemblage of subcritical shapes and pieces, and suddenly reassembling it in a single, dense, supercritical piece. The Hiroshima bomb involved the so-called gun design, in which a gun-like device fired a U-235 projectile into a cavity in another piece of the same material; separately they were subcritical, together they were supercritical. For reasons beyond the scope of this book, this method does not work with plutonium. In that case, the so-called implosion method has to be used. In that method, a subcritical assemblage of plutonium is completely surrounded by a large mass of high explosive, the high explosive is ignited at many points on its outside surface simultaneously, and a powerful shock wave converges inwards on the plutonium. The resulting "implosion" of the plutonium causes it to be suddenly reformed into a dense, supercritical mass that promptly produces a violent nuclear explosion which exceeds by thousands of times the implosion needed to initiate it.

they came. Many of them had been able to find occasional chinks of time to think about what they would do "after the war" and had generated some ideas they were especially anxious to explore further.

Robert Oppenheimer was among those who left Los Alamos with the intention of resuming a career in the academic world. At first he divided his time between two simultaneous appointments as Professor of Theoretical Physics at the University of California in Berkeley and at the California Institute of Technology in Pasadena, but shortly after (1947) he left California altogether to become director of the Institute for Advanced Study at Princeton, New Jersey. In addition, and much more important in the present context, from the end of the war until 1952 Oppenheimer spent a very large part of his time and nervous energy on a nominally part-time basis helping to work out the details of America's postwar nuclear programs and policies. He was one of the principal authors of the so-called Baruch plan, he was chairman of the principal scientific advisory committee to the newly created AEC, and he served on many other standing and *ad hoc* committees in other branches of the government, including the Department of Defense, the State Department, and the White House. This remarkable web of interconnected advisory posts, in combination with his great intellectual power and the special mystique that surrounded him as a result of his having directed the central part of the atomic bomb program, made him by far the most influential nuclear scientist in America during the immediate postwar period.

Some members of the wartime Los Alamos staff, for various reasons, chose to stay, or, at least, were willing to give serious consideration to staying. Some of them liked the location on the high bright desert mesa, others were intellectually interested in the problems they were working on and wanted to see some of the investigations and ideas carried further, and still others responded to the first chill winds of the cold war and were motivated to continue their work on atomic weapons for that reason. And no doubt some others, including those who had been caught up in work at Los Alamos as their first professional job after leaving graduate school, simply did not have a good "home" to go back to and did not receive any interesting offers of new employment.

The situation at Los Alamos at war's end was further complicated by a pervasive air of uncertainty as to the intentions of the government. It was clear that the overall national management of the program would change hands, but there was much controversy over the form that the new management would take. Indeed, it was to take sixteen months from the end of the war until the newly created civilian AEC took over from the army. In the meantime General Leslie R. Groves did his best to assure those who were willing to consider staying that the laboratory and its program had a future, but it was clear to all that he was only expressing his personal views.

In view of all this, it is remarkable that the laboratory survived, and most of the credit must go to Oppenheimer's successor: Norris E. Bradbury. Bradbury was a Berkeley Ph.D. with some experience as a physics professor at Stanford before the war. He was assigned (as a naval officer) to Los Alamos in July 1944. On Oppenheimer's advice, Groves appointed him director of the laboratory in October 1945, and he continued in that post until his retirement in 1970.

Edward Teller, who had been one of the members of the Theoretical Division during the war, was among those who seriously considered staying, and Bradbury invited him to take over the direction of the Theoretical Division on a permanent basis. In response, Teller proposed that there be twelve[2] nuclear weapons tests per year and made Bradbury's acceptance of his proposal a condition of his staying. Bradbury regarded that testing rate as entirely unrealistic and Teller left Los Alamos for the University of Chicago, where a number of other wartime intellectual leaders also gathered. As it turned out, test rates did not regularly reach a level of twelve per year until 1955 in the United States, by which time there were two active nuclear weapons laboratories, and they did not reach this level in the U.S.S.R. until 1957. No other country has ever tested more than a few each year.

The laboratory recovered slowly but steadily from the postwar exodus. Bradbury reported[3] that there were only eight theoretical physicists on the staff in 1946. This number increased to twelve by the end of 1947, fourteen in 1948, twenty-two in 1949, thirty-eight in 1950 (after the H-bomb acceleration), and fifty in 1953.

The efforts of this small group were, however, augmented by frequent visits, sometimes lasting for months, by many of the wartime leaders, including Teller, Bethe, Von Neumann, and Fermi.

The total laboratory staff during this early postwar period totaled about 1200 persons.[3] The laboratory not only carried out research and development whose purpose was to improve the atomic bombs but also continued for several years to refine plutonium and uranium, to produce high explosive detonators, and to participate in various other manufacturing activities.

The postwar weapons research program of the laboratory had two main objectives. One objective was the improvement of atomic bombs, or fission bombs as they might be more correctly labeled. The other was the exploration of an entirely different type of nuclear energy based on fusion processes, also known as thermonuclear reactions. It is this latter process that provides the energy of the superbomb.

The objective occupying by far the largest part of the staff was the improvement of fission bombs. The goals at first were not very imaginative; they were mainly concerned with improving reliability and safety, and increasing the efficiency of fissile materials by further development of the implosion method. And this was a satisfactory program from the point of view of the military users who continued to think only in terms of delivery by large strategic aircraft such as the B-29 that carried the bombs to Hiroshima and Nagasaki. It was only several years later that they began to think seriously of delivery by other means, such as missiles, torpedos, and artillery, which created firm requirements for greatly reduced dimensions and weights. Finally, in 1948, the laboratory conducted the first weapons tests whose purpose was to improve on the design of the wartime bomb.[4] This test series,

---

[3]This number does not include housekeeping staff and guards, who were hired by the Zia Corporation. It is therefore not directly comparable with figures for other laboratories, which usually include such persons.

[4]An early operation, called Crossroads, was conducted at Bikini in 1946, but its objective was the evaluation of the effects of existing bombs rather than their improvement.

known as Operation Sandstone, was conducted by Joint Task Force Seven in April and May at Eniwetok Atoll, whose natives had been evacuated for the purpose. Since in addition to the AEC's interest in improved weapons designs there was a strong military interest in weapons effects and since a very large logistic effort was required in connection with such an overseas operation, a joint AEC–Defense Department organization, known as Joint Task Force Seven, was formed for the purpose. The task force commander was General John E. Hull, the test director was Navy Captain James S. Russell, and the scientific director was Darol Froman of the Los Alamos staff. There were three test explosions, ranging in yields from 18 to 49 kilotons.[5] The tests evidently did result in substantial improvements in the efficiency of use of fissile material. By making use of new designs based on these experiments, the commission reported, "[it] would now be able to produce more weapons than had been required in the schedule which the Joint Chiefs of Staff had prepared late in 1947[4]."

The success of the Sandstone tests also boosted morale at Los Alamos and helped to garner further support for the laboratory in Washington. As a result, the construction of a new laboratory, located nearby on South Mesa, was authorized as a replacement for the wartime facilities that were still being used. The test operation was also a success from an administrative point of view, and the mixed military–civilian, joint-task-force style of operation became a permanent method of conducting overseas atomic weapons tests.

Work on the Super, 1942–1949

In addition to research intended to improve fission weapons, the Los Alamos laboratory throughout its entire history has also conducted research on methods for releasing energy through the fusion process. In this process, the nuclei of two very light atoms (usually isotopes of hydrogen) combine to form a heavier nucleus (usually a helium nucleus). It is this combining, or fusing, of nuclei that

[5]The Hiroshima explosion was 13 kilotons and the Nagasaki explosion was 21 kilotons.

gives the process its name. Because this particular nuclear process will only take place at extraordinarily high temperatures—tens of millions of degrees—it is also called a thermonuclear reaction, and because the nuclear weapons that derive most of their energy from fusion processes use various forms of hydrogen as their fuel, they are frequently called hydrogen bombs.

Early in this century it was recognized that the stars—including the sun—probably derived their enormous output of energy from some nuclear process. In the thirties, Hans Bethe worked out the details of just how this might actually come about through a somewhat complex set of thermonuclear processes which could take place under the conditions of exceedingly great pressures and enormously high temperature believed to prevail in the center of stars.[6] These conditions were so utterly different from any yet attained or then projected on earth, that no one at the time thought seriously about producing such reactions here. However, with the advent of the atomic bomb, the prospects were dramatically altered. In the center of an exploding fission bomb, temperatures substantially exceeding 100,000,000 degrees are produced, and so at least one of the conditions necessary for igniting a thermonuclear reaction under the control of man seemed to be within reach.

According to Edward Teller, Enrico Fermi in early 1942 asked him, "Now that we have a good prospect of developing an atomic bomb, couldn't such an explosion be used to start something similar to the reactions in the Sun?[5]." Specifically, Fermi had in mind reactions involving deuterons, the nuclei of a relatively rare but naturally occurring heavy form of hydrogen. As we now know, and as Fermi then speculated, at temperatures of the order of one hundred million degrees (actually much hotter than the interior of the sun) deuterons react explosively with each other in the fusion process, producing helium and huge amounts of energy. Compared to ordinary chemical processes such as burning oil or coal, the amount of energy produced by an equal weight of reactants is about ten million times larger. And even compared to the process of uranium fission, the energy per unit weight is

[6] In 1967, he was awarded the Nobel Prize in physics for this work.

nearly three times greater. Teller reports that "after a few weeks of hard thought, I decided that deuterium could not be ignited by atomic bombs." Shortly after, in the summer of that same year, Robert Oppenheimer gathered a small group[7] of theoretical physicists at Berkeley to discuss atomic bombs and to lay out a program for the projected new laboratory at Los Alamos. By this time, Teller's views had shifted, and he and Emil Konopinski presented more optimistic calculations to this group. The group then discussed the hydrogen bomb at length, along with many other questions concerning the uranium bomb. During these discussions the idea of using tritium as well as deuterium was first suggested. Tritium is a still heavier form of hydrogen (having three times the normal weight) which does not normally occur in nature. It was well known, however, that tritium could be produced artificially by causing neutrons to react with lithium. We now know that a fifty/fifty mixture of tritium and deuterium reacts about one hundred times as rapidly as does pure deuterium.

Thus, when Los Alamos was established, the exploration of the super was among the original objectives. However, because the development of fission bombs turned out to be more difficult than expected, their development demanded and received virtually the full attention of the laboratory. Only a relatively small group, under Teller's direction, put in much effort on the super during the war.

In the spring of 1946, a group of those who had remained at Los Alamos after the postwar exodus, augmented by some visitors from among the wartime group, again took up the study of how thermonuclear reactions might be produced on the earth. This study soon branched along two quite distinct lines with very different basic objectives. One such line of research had the comparatively easy objective of igniting a relatively small mass of thermonuclear fuel by means of the energy produced in a relatively large fission explosion. As we shall see in chapter 5, the United States successfully accomplished this objective in 1951, and the Soviet Union did so in 1953. This particular objective later became

[7]Other members of the group were Hans Bethe, Stanley Frankel, Emil Konopinski, Eldred Nelson, Robert Serber, and J. Van Vleck.

important in connection with a process known as "boosting" or the "booster principle." These terms "refer to the notion of using a fission bomb to initiate a small thermonuclear reaction with the possibility that . . . the neutrons from this reaction might increase the efficiency of the use of the fissile material."[8] This means that in certain circumstances there can be a synergistic interaction between the fission and fusion reactions that can substantially increase the efficiency of the fission reaction.

The other line of thermonuclear research had the very much more difficult objective of igniting a very large—in fact, indefinitely large—mass of thermonuclear fuel by means of a relatively small fission explosion. From the beginning, the goal of this line of research was the design of a superbomb that would yield, in round figures, an explosion a thousand times as large as that of a nominal fission bomb. The United States produced an explosion of this much more difficult type in 1952, and the Soviet Union did so in 1955.

A report on the status of our understanding of thermonuclear processes as of the spring of 1946 was issued on June 12, 1946. It was entitled "Report of Conference on the Super." A very heavily censored version of this report was declassified in 1971 for use in connection with litigation concerning priorities in the development of electronic computers[6].

Among the thirty-one persons who were listed as having participated in the conference, some, such as Emil Konopinski, Lothar Nordheim, George Placzek, Robert Serber, Edward Teller, and John Von Neumann, were already widely recognized for their

---

[8]The authors of a summary report of the work done on the super during the war were E. Teller, E. Konopinski, S. Frankel, H. Hurwitz, R. Landshoff, N. Metropolis, and A. Turkevich. Boosting was mentioned in passing in L. L. Strauss, *Men and Decisions*, Garden City, N.Y.: Doubleday & Company, and alluded to in a press conference by Norris Bradbury given at Los Alamos in 1954. Boosting and the research behind it have recently been discussed in some detail in Carson Mark, *A Short Account of Los Alamos Theoretical Work on Thermonuclear Weapons, 1946–1950*, LA-5647-MS, Los Alamos, N. Mex., 1974. This last report also mentions another 1946 idea known as the TX-14. This is another approach to the problem of igniting very large masses of thermonuclear fuel. Since there is no way to make clear here how its details differ from those of the so-called classical super, and since there are no political or strategic distinctions among these two and yet other variants of superbombs, I ignore it here.

work in basic science.[9] Three members of the wartime British "team" at Los Alamos also attended, including Dr. Emil Klaus Fuchs who, as it was later learned, was passing on what he knew to the Russians. The only participant from the armed services was Colonel Austin R. Betts, who later became director of the AEC's Division of Military Applications (1960–1964). Of all the attendees, only five continued on at Los Alamos as regular full time members of the Theoretical Division's staff. These were Rolf Landshoff, Carson Mark, Frederick J. Reines, R. D. Richtmyer, and Stanislaw Ulam.

The report concluded that "the detailed design submitted to the conference was judged on the whole workable. In a few points doubts have arisen concerning components of this design. These doubts have been discussed above. In each case, it was seen that should the doubts prove well-founded, simple modifications of the design will render the model feasible." As we now know, the optimism expressed in their conclusion was unwarranted, and the doubts that were raised turned out to be very well-founded indeed. Overcoming the difficulties behind these doubts required a new invention, which did not come along until another five years had passed, rather than the "simple modifications of the design" that the report predicted.

At the end of the report, the conferees noted that "The undertaking of the new and important Super Bomb Project would necessarily involve a considerable fraction of the resources which are likely to be devoted to work on atomic developments in the next years. A statement of the potentialities of the super as a weapon, and an estimate of the cost of answering, in practice, the questions still unsolved, have been included. But we feel it appropriate to point out that further decision in a matter so filled with the most serious implications as is this one can properly be taken only as part of the highest national policy."

For the next several years, work on the super progressed rather slowly, and involved a relatively small number of persons. Al-

[9] It is interesting to note that one of these men, Placzek, had visited the U.S.S.R. in the late thirties to ascertain whether or not central European refugee nuclear physicists could find suitable positions there. He is reported to have found the atmosphere most uncongenial.

though as reported above the number of fully qualified physicists and mathematicians in the theoretical group increased steadily after the middle of 1946, it was sometime in 1949 before there were even as many as twenty; and only a portion[10] of their effort could be reserved exclusively for studying the super. This small group carried out calculations designed to elucidate the behavior of the existing theoretical designs and at the same time tried to discover more promising variations. In addition to this central theoretical work, experimental measurements were made in order to determine the details of the various reactions involved. During those same years, most of the experimental scientists and engineers at the laboratory were engaged in work intended to produce further improvements in and better understanding of fission bombs. The work on fission was, of course, useful not only as an end in itself; it was also an essential preliminary to the later successful development of the hydrogen bomb.

There were other more fundamental reasons the work on the superbomb went slowly during the 1946–1948 period. The ideas that did exist, the so-called classical super and the alarm clock[7], were, as Oppenheimer later put it, "singularly proof against any form of experimental approach[8]." In the case of fission bombs, there were several good methods available for investigating the bulk behavior of fissile material on a laboratory scale. These included observations of the properties of fast reactors and just momentarily critical bomb-like assemblies, and highly instrumented experiments in which ordinary uranium was imploded by chemical explosives. Such laboratory observations, when combined with a growing body of theory, made it possible to predict fairly reliably the performance of a new fission bomb before it was actually built and tested. In the case of thermonuclear bombs, there were no known ways to investigate the reaction process in bulk in the laboratory. The only way to study the fusion process in even a small mass of fuel was to subject it to the extreme heat and enormous energy output of a full-scale nuclear explosion. Such experiments are very difficult and expensive, and for that

[10]Carson Mark, director of the LASL Theoretical Division from then until 1973, estimates that nearly one half of the division's effort was devoted to the super in the 1946–1949 period.

reason Bradbury insisted that the theorists should have some reasonably clear ideas of where they were going before any such experiments were planned. Nowadays, theoretical calculations conducted with the aid of high-speed electronic computers provide an effective alternative to experiments, but the very small capacity of such machines as were then available did not permit calculations using mathematical models that were anywhere near realistic. Consequently, they always produced ambiguous and uncertain results.

In the meantime, work on the reverse state of affairs, that is, on the problem of how to ignite a small amount of thermonuclear fuel by means of a large fission explosion, was making headway. One particular application of this work was called the "booster principle," a means of increasing the efficiency of fission weapons already mentioned above. According to Carson Mark, such possibilities were recognized at least as early as 1946. In the summer of 1948 a detailed study of a design incorporating the boosting principle was made, and a full-scale test of a device of this type was put on the list of test explosions to be made in the next overseas test operation, then already being planned for 1951. These studies were carried out by Marshall Rosenbluth and John Reitz under the general direction of Teller. By the fall of 1948, the most promising approaches to the problem had been identified, and almost all aspects of the problem had been studied before the first Russian test in 1949. In January 1950, a specific experimental model was selected, and in late October the last details of the design were frozen. It had been recognized all along that a test of the booster type of device was not only interesting as an end in itself, but that it would also provide an opportunity to explore experimentally some of the key phenomena involved in thermonuclear burning.

By mid-1949, work on the super itself presented a very mixed picture. On the one hand, a steadily accumulating body of theoretical calculations, made using such methods and data as were available, began to make it appear likely that the classical super of the early forties either would not work at all, or if it could be made to work it would require very large amounts of the extremely expensive artifically produced hydrogen isotope tritium. There was, moreover, at that time no really promising alternative.

On the other hand, by that time the details of the deuteron–deuteron and the deuteron–triton reactions were becoming better known, fairly good estimates of the reaction rate as a function of temperature and density were possible, and the various mechanisms for heating and cooling a reacting mass of light elements were becoming well understood. In addition, it was realized that several different thermonuclear fuel combinations were possible: pure deuterium, deuterium plus varying amounts of tritium, and lithium deuteride in which the lithium was either the naturally occurring form, or was lithium especially enriched in the light isotope of atomic weight six (Li^6).[11] Moreover, computing devices and computing techniques advanced to the point where it began to appear to be possible to make more definitive calculations of different kinds of thermonuclear devices, and there had been further progress in the theory and design of fission weapons.

The net situation was, as Teller summarized somewhat later when he was trying to enlist support for a high-priority program, one in which "We still don't know if the Super can be built, but now we don't know it on much better grounds[9]."

The result was that a number of theorists outside Los Alamos, Teller among them, found the prospects for further work on the super at Los Alamos sufficiently attractive and promising to cause them to join on a full-time basis the small group already working there under the direction of Carson Mark, the Theoretical Division leader. (Teller, Von Neumann, and some others from the wartime staff had continued to participate in the Los Alamos laboratory program by consulting two to three months in each of the intervening years.) Thus, when the first Soviet explosion in August 1949 ("Joe 1") provided a powerful political stimulus to expand and accelerate the thermonuclear program, the situation was ripe for doing so.

Whether or not the effort put into the superbomb during the 1946–1949 period was adequate and appropriate was the subject of a very acrimonious debate that went on for many years thereafter. Teller, who always contended it was not, wrote in 1962 that "If the Los Alamos Laboratory had continued to function after Hiroshima with a full complement of such brilliant people as Oppenheimer, Fermi and Bethe, I am convinced that someone would have had the same idea much sooner—and we would have had the hydrogen bomb in 1947 instead of 1952[10]." Norris Bradbury, the laboratory director from 1945 to 1970, held the contrary view. In 1954 he said,

> We would have spent time lashing about in a field in which we were not equipped to do adequate computational work. We would have spent time exploring by inadequate methods a system which was far from certain to be successful—I cannot see how we could have reached our present objectives in a more rapid fashion [than] the mechanism by which we went[11].

We will return to this particular point later.

---

[11]Tritium is normally produced by the reaction neutron plus lithium six yields tritium plus helium, $Li^6 + N \rightarrow H^3 + He^4$. There are two possible ways to bring about this reaction on a large scale. One is to cause it to take place in a reactor, in which case the tritium is later chemically separated from the lithium and mixed with deuterium in the required concentration. Alternatively, the reaction can be caused to take place *in situ* in a hydrogen bomb using lithium deuteride (a grayish saltlike solid) as its fuel. In that case, when two deuterons react, they can produce a neutron, and that neutron can in turn react with lithium six to produce a triton. The triton then very quickly reacts with another deuteron and in that reaction produces yet another neutron which can repeat the cycle. This cycle can also proceed along several alternative routes. For instance, the "first" neutron could have originated in a nearby fission event in uranium or plutonium. In another alternative, the very high energy neutron produced in the triton-deuteron reaction can readily cause fission even in U-238 if any happens to be nearby. (The use of LiD in thermonuclear weapons is discussed by John Foster in the *Encyclopedia Americana,* 1971, Vol. 20, p. 522.) The basic idea of using this complex chain-like process was, to my knowledge, already known at Los Alamos in the summer of 1947, and may well have come up at least a year before that. Natural lithium contains only 7.4% Li^6. If the proportion of this isotope is increased by artificial means, the whole process obviously goes much faster. However, such enrichment is, by normal chemical industry standards, very expensive, and like other large-scale isotope separation processes, it requires a very substantial initial capital investment.

Chapter THREE

The Soviet Nuclear Program to 1949

The announcement of the discovery of fission in Germany in 1939 caused great excitement among nuclear physicists in the U.S.S.R., just as it did in all Western countries. Soviet work in nuclear physics[1] was on a more modest scale than in Germany, England, France, and the United States, but the quality of the work that was done was comparable. And while direct personal contacts between Soviet and other scientists were severely limited, the circulation of scientific journals from outside was not seriously impeded and the significance of the new discoveries was recognized as quickly there as elsewhere. It is reported that in 1939 Igor Tamm, a leading Soviet physicist, remarked to a group of students, "Do you know what this new discovery means? It means a bomb can be built that will destroy a city out to a radius of maybe ten kilometers[2]." As elsewhere, physicists in Russia began to make measurements and calculations intended to elucidate the new discoveries and to determine under exactly what conditions, if any, a chain reaction would take place. One product of Soviet research on this subject was the discovery of spontaneous fission, an exceedingly rare event, by G. N. Flerov and Petrzhak,[1] working under the general direction of Igor Kurchatov. A cablegram describing this discovery was sent to the American scientific journal *Physical Review*, which published it in its July 1, 1940, issue (p. 89). According to Igor Golovin[3] (a professional colleague of Kurchatov who authored an excellent biography of the latter), the complete lack of any American response to the publication of this discovery was one of the factors which convinced the Russians that there must be a big secret project under way in the United States.

The invasion of the Soviet Union by the Nazis in June 1941 brought most work in nuclear physics to a halt as scientists and technicians were drafted into the army or mobilized to do other work of more immediate value, such as the development of radar and making ships safe against magnetic mines.

But work on uranium was not entirely forgotten. At an international "anti-Fascist meeting" of scientists in Moscow, on October 12, 1941, Peter L. Kapitza, one of the leaders of Soviet physics, in commenting on what scientists could do to help the war effort said in part, "In recent years a new possibility—nuclear energy—has been discovered. Theoretical calculations show that, if a contemporary bomb can for example destroy a whole city block, an atomic bomb, even of small dimensions, if it can be realized, can easily annihilate a great capital city having a few million inhabitants[4]." Some months later Flerov, who had been following the foreign literature on the subject, wrote to the State Defense Committee, "no time must be lost in making a uranium bomb." He wrote similarly to Kurchatov, explaining the reasoning behind his conclusion. In the meantime, according to Golovin, the Soviet government had come into possession of information showing that urgent top secret work on atomic bombs was in progress in both the United States and Germany. The government turned to a group including Abram Joffe and Peter Kapitza to ask whether the

[1] In 1946 Flerov and Petrzhak were awarded the Stalin Prize for this discovery.

U.S.S.R. could and should also mount a program in the field. The net result was the initiation of a program under the direction of Kurchatov in February 1943.[2]

The Soviet A-Bomb

During the war years, the Soviet program did not come close to matching the U.S. Manhattan project. The number of physicists involved was said to be about twenty, and the total staff at the main laboratory was only fifty persons altogether. Even so, they did experiments and made theoretical calculations concerning the reactions involved in both nuclear weapons and nuclear reactors, they began work designed to lead to the production of suitably pure uranium and graphite, and they studied various possible means for the separation of the uranium isotopes.

At the Potsdam Conference, in July 1945, Truman told Joseph Stalin about the U.S. A-bomb program for the first time. According to Truman[5], "I casually mentioned to Stalin that we had a new weapon of unusual destructive force. The Russian Premier showed no special interest. All he said was he was glad to hear it and hoped we would make good use of it against the Japanese."

Soviet Marshall Zhukov later wrote that after Truman's casual announcement Stalin called him and Soviet Foreign Minister V. M. Molotov aside and said, "They simply want to raise the price. We've got to work on Kurchatov and hurry things up[6]." Apparently the word was passed at once to Kurchatov, for Golovin confirms that Kurchatov and his closest associates did learn of the Alamogordo test before the Hiroshima attack took place.

The explosion of the three American atomic bombs in close succession removed all the lingering doubts about the feasibility and desirability of building nuclear weapons and was, without doubt, the most important intelligence datum in the possession of the Soviets. The Smyth report[7], which described the U.S.

[2]He was joined by I. K. Kikoin, Ya. B. Zeldovitch, A. I. Alikhanov, G. N. Flerov, Yu. Ya. Pomeranchuk, B. V. Kurchatov (Igor Kurchatov's brother), I. I. Gurevich, G. Ya. Schchepkin, Yu. B. Khariton, M. S. Kozodaev, V. P. Dzhelov, L. M. Nemenov, and V. A. Davidendov.

program, was issued in the United States within days after the first explosions over Japan and was very widely circulated—thirty thousand copies in the first Russian printing—in the U.S.S.R. very soon thereafter.

As a result the Soviet government immediately took measures designed to expand and accelerate their efforts. An engineering council was set up under the chairmanship of Boris L. Vannikov. Vannikov, whose role Arnold Kramish (an American authority on the early Soviet program) compares with that of General Groves, was a Soviet army general who had been People's Commissar of Munitions during the war. Kurchatov and M. G. Pervukhin, commissar of the Chemical Industry, were appointed Vannikov's deputies. The council was made up of engineers, industrial managers, and scientists, including A. I. Alikhanov, I.K. Kikoin, A. P. Vinovgradov, Abram Joffe, and A. A. Bochvar. Another man who was to play a major role was Colonel General Avraamy Zavenyagin of the NKVD (the secret police). He, interestingly enough, was not only involved with administration and security, but evidently pitched in on the technical work itself, apparently being directly involved in the preparation and final assembly of the first Soviet bomb. Many of the problems the Russians faced and overcame were the same as those which the American project had faced earlier: production of adequately pure uranium and graphite in sufficient amounts, unexpected differences between the behavior of large high-power production reactors and the smaller pilot plant version, determining the chemical and physical properties of plutonium from exceedingly small samples—first produced in a cyclotron and then later by the first low-power graphite uranium reactor, the production of U-235 by gaseous diffusion, and the design of a chemical explosives system for imploding plutonium in order to achieve a supercritical mass. Among those who worked on the industrial problem involved was Vassily Emelyanov, a metallurgist who built tanks during the war, and who later (in the fifties and sixties) very frequently represented the Soviet Union in international atomic circles.

The Russians first achieved a chain reaction on Christmas Day 1946, in a graphite uranium lattice that apparently was much like Fermi's reactor in Chicago. That reactor, known as the F-1, apparently still exists today on the site where Kurchatov and his colleagues put it together more than a quarter of a century ago[8]. The first production reactor apparently began to work satisfactorily in the fall of 1948 after first creating some difficult problems for the designers that may have been much like those that the United States had to overcome during the war. Finally, on August 29, 1949, the Soviets exploded their first atomic bomb in Asia near Semipalatinsk. The following excerpt from Golovin gives the Soviets' own view of this event and their perceptions about how it fitted into the general scheme of things:

> The creators of all the weapons components gathered on the test site. With their own hands, the directors—physicists and engineers—brought the parts up to the requisite level of reliability. Work went on around the clock.
>
> Kurchatov and Zavenyagin personally directed all the preparations on which the success of the test depended. Both could sometimes be seen on the spot of the future explosion, sometimes where the bomb components were being assembled, and sometimes in the concrete laboratory bunkers.
>
> At last the bomb was assembled under the tireless direction of Kurchatov and Zavenyagin. Finally, it was lifted up onto the metal tower where it was to be exploded.
>
> The test was successfully carried out on August 29, 1949 in the presence of the Supreme Command of the Soviet Army and Government leaders.
>
> When the physicists who had created the bomb saw the blinding flash, brighter than the brightest sunny day, and the mushroom cloud rising into the stratosphere, they gave a sigh of relief. They carried out their duties. No one became frightened like the physicists in the U.S.A. who had gathered from everywhere and had made the weapon for the army of a country that was foreign to many of them and whose government used it against the peaceful populations of Hiroshima and Nagasaki.
>
> The Soviet physicists knew they had created the weapon for their own people and for their own army which was defending peace. Their labor, their sleepless nights, and the huge effort that had constantly increased in the course of those past years had not been in vain: they had knocked the trump card from the hands of the American atomic diplomats[9].

Within days, the radioactive debris from the explosion was picked up by U.S. aircraft equipped specifically for the purpose. These aircraft were available and on station largely because of special efforts by Lewis Strauss, then a member of the AEC, and especially sensitive to and worried about what the Soviets might be doing. On September 23, President Truman announced the explosion of the bomb that came to be known as "Joe 1" (for Joseph Stalin):

> We have evidence that within recent weeks an atomic explosion occurred in the U.S.S.R.
>
> Ever since atomic energy was first released to man, the eventual development of this new force by other nations was to be expected. This probability has always been taken into account by us. . . .
>
> The recent development emphasizes once again, if such emphasis were needed, the necessity for that truly effective, and enforceable international control of atomic energy which this government and a large majority of the United Nations support.

(Truman, as he said later, simply could not bring himself to believe "those asiatics" could built something as complicated as an atomic bomb. He made David Lilienthal, Robert F. Bacher, and the other members of the special committee appointed to study the evidence, personally and individually sign a statement to the effect they really believed the Russians had done it[10].)

On September 25, 1949, the Soviet news agency TASS replied to Truman's announcement and other stories in the Western press with a classic statement that typifies the secrecy and obfuscation with which the Soviets cloaked even the obvious in the last Stalin years:

> In the Soviet Union, as is known, building work on a large scale is in progress—the building of hydroelectric stations, mines, canals, roads, which evoke the necessity of large-scale blasting work with the use of the latest technical means.
>
> Insofar as this blasting work has taken place frequently in various parts of the country, it is possible this might draw attention beyond the borders of the Soviet Union.
>
> As for the production of atomic energy, TASS considers it necessary to recall that already on November 7, 1947, minister of Foreign

> Affairs of the U.S.S.R. V. M. Molotov made a statement concerning the secret of the atom bomb, when he declared that this secret was already long ago nonexistent.
>
> This statement signified the Soviet Union already had discovered the secret of the atomic weapon and that it had at its disposal this weapon. . . .
>
> As for the alarm that is being spread on this account by certain foreign circles, these are not the slightest grounds for alarm. . . . The Soviet Government, despite the existence in its country of an atomic weapon, adopts and intends adopting in the future its former position in favor of the absolute prohibition of the use of the atomic weapons, and control will be essential in order to check up on fulfillment of a decision on the prohibition of production.

With official statements like that, it is no wonder that wild rumors circulated and were believed.

It is interesting to compare how long it took the Russians to make their first A-bomb with the *a priori* estimates of how long it would take. A majority of the estimates made at the end of the war, particularly those made by scientists familiar with the program, said that it would take the Russians "about five years." These estimates were, giving due regard to the inherent difficulties in making them, remarkably close. For example, Hans Bethe and Fred Seitz, in their essay in the collection *One World or None* published in the fall of 1946, said that "we are led by quite straightforward reasoning to the conclusion that any one of several determined nations could duplicate our work in a period of about five years[11]." Vannevar Bush is reported to have told Navy Secretary James V. Forrestal in September 1945 that if the Russians concentrated their resources they might equal our 1945 position by 1950[12]. A group at the Bell Laboratories, headed by William O. Baker, was asked by the State Department to make a careful study of the question and they, too, came up with a five-year estimate[13]. And perhaps most interesting of all, Golovin says that when the Soviet government asked Kurchatov how long it would take "if he received all-sided support," he responded, "five years[14]."

Some estimates were higher and others lower. Groves is reported to have estimated it would take the Russians as long as twenty years. Winston Churchill[15] said the U.S. monopoly would last only three or four years and so did the authors of the Franck report[16]. In general, scientists favored lower estimates, and administrators and politicians (except Churchill!) leaned toward higher ones. Contemporary polls show that a majority of the general public accepted the scientists' assessment[17].

The timing of the Soviet program was remarkably similar to the U.S. program's. From initiation (that is, August 1945) to completion it took them four years exactly; only five months more than in our case. (The period from Vannevar Bush's recommendation for a crash program, which was made the day before the attack on Pearl Harbor, December 6, 1941, to the test on July 16, 1945, was three years and seven months.) Even more remarkable was the time it took to go from the first chain reacting pile (December 2, 1942, in our case; December 25, 1946, in theirs) to a first bomb test. The periods in the two cases differ by only eighteen days out of a total time of two and a half years!

In retrospect it appears that two important differences worked in opposite directions and compensated for each other. The United States had a very substantial quantitative advantage in human and material resources; the Soviet Union had the advantage of knowing the bomb was feasible (from the Alamogordo test and the atomic attacks on Japan) and that any of several approaches to its manufacture would work (from the Smyth report).

The Role of Espionage

A question is often raised about the role of espionage in the development of the Soviet atomic bomb.

When it was discovered in 1950 that Klaus Fuchs, who had been a member of the British team at Los Alamos during the war, had all along been a Soviet spy, it was quite logically concluded that espionage had played a major role in the Soviet development. There had been an earlier revelation of a Soviet atomic spy ring operating in Canada during the war, but only peripheral matters seemed to have been involved. Fuchs, on the other hand, had been at the very center of things, and had apparently passed on much of what he knew, including some of the earliest ideas about the hydrogen bomb. For that reason, press stories about the Fuchs case widely credited Fuchs' treachery with cutting years off the Soviet program. And the congressional Joint Committee on Atomic Energy (JCAE) reported that "Fuchs alone has influenced the safety of more people and accomplished greater damage than any other spy not only in the history of the United States but in the history of nations[18]."

On the other hand, the fact that the Soviet program took about as long to accomplish (four years) as most well-informed *a priori* estimates said it might ("about five years"), and the fact that the Soviet program as described by Golovin had almost the same spacing of events as ours did, and that it included research programs that could have produced all the necessary data, argue that espionage did not play an essential role.

Unfortunately, there is as yet not enough information to form a certain judgment one way or the other.

Peter Kapitza and the Bomb

In the mid-1950's, rumors began to circulate to the effect that Peter Kapitza, one of the most distinguished of all Soviet scientists, had earlier refused to work in the atomic bomb program, and had, as a consequence, been arrested. The rumors have turned out to be true, and the episode provides a most interesting sidelight on Soviet scientific affairs.

Peter Kapitza is a physicist, and today one of the most senior and most honored members of the prestigious Soviet Academy of Science. As a young man, he worked in nuclear physics in England under the great Ernest Rutherford's tutelage at the Cavendish laboratory. For a time he was able to travel back and forth between England and the U.S.S.R., but during a home visit in the mid-1930's, and quite unexpectedly, he was refused permission to leave Russia again. He then successfully built up a laboratory in Moscow in which he did research involving extremely high magnetic fields and extremely low temperatures. For years his laboratory had the best facilities in the world for producing liquid hydrogen and liquid helium.

During the war he served on state and academy committees which advised the government on many scientific questions including those related to the atomic bomb, but he was not a member of the small group that actually worked directly on it.

After the explosion of the three American bombs in July and August of 1945, the Soviet program was greatly accelerated. In 1946, it was placed under the direct authority of Lavrenti Beria, chief of the secret police. It was soon after this that Kapitza "refused" to do further work on the bomb. As he assured me, his refusal was not based on "political or moral reasons." He said he believed that "since your country had the bomb, my country would have to have it too." As far as I know, no one in the West knows the detailed story of just what did happen, but the following is what I believe happened, and is based partly on Kramish's[19] speculations about the matter and partly on some conversations with a number of other persons, each of whom knows at least a part of the truth.

Apparently, not long after Beria took charge of the program, a serious argument between Beria and Kapitza took place. The disagreement was over an essentially technical issue, and seems to have involved some differences over the appropriate role of Kapitza's laboratory. Beria thought the laboratory should be doing one thing while Kapitza thought it should be doing something else, perhaps not even directly related to atomic energy.[3] Each apparently stuck to his position, and Beria, who was of course very powerful, had Kapitza placed under house arrest in his dacha outside Moscow (his laboratory and his regular home are in the city). There he remained until after the death of Stalin, cut off from his work and his friends, save for a few who risked seeing him from time to time. During that period, he apparently did not suffer in the physical sense; his family visited him and his basic salary as an academician seems to have been continued,

[3] Kramish suggests that the argument involved the separation of uranium isotopes. I have also heard that it involved the production of deuterium. Also, in the most recent of the so-called Khrusnchev Memoirs (in *Time* magazine, May 6, 1974, pp. 38–45), there is what appears to be a badly confused reference to this same matter. In it, it is evident Khrushchev has no understanding of Kapitza's point of view, and puts the whole thing down as merely a selfish desire for fame on the latter's part.

but the political climate was fearsome, and the psychological stress must have been considerable. Even so, during the period of house arrest he somehow managed to publish a paper on the motion of a pendulum with a vibrating support, and to do some research on ball lightning[20].

Early in Nikita Khrushchev's regime, Kapitza was restored to his former position, and he returned to his laboratory and his city home. Since then he has again been able to travel widely and to participate in international scientific activities, including the politically oriented Pugwash conferences. In the fall of 1973, Kapitza was one of the few—perhaps the only—academicians who were asked to sign the public statement[4] attacking Andrei Sakharov but who declined to do so. (Sakharov, also an academician, had been one of the leaders of the Soviet H-bomb program in the fifties. In the late sixties and the seventies, he was one of the small band of dissident Soviet intellectuals that publicly protested certain repressive acts of the government.) Even so, in the summer of 1974 on the occasion of his eightieth birthday, he was awarded his second "Hero of Socialist Labor." (The first was in 1945.)

All of this is seen to be especially ironic when we note that during the debate in the United States over the hydrogen bomb, which took place while Kapitza was under house arrest, Kapitza's knowledge of nuclear physics and his experience in producing large quantities of liquid hydrogen were sometimes cited as indications of probable Soviet progress towards an H-bomb[21].

But this matter is not only ironic, it is instructive as well. Among those who cited Kapitza's possible leadership role in the Soviet thermonuclear program was Lewis Strauss. It must be presumed that Strauss had good access to what American intelligence data there were on the subject, and that he would have known Kapitza was under house arrest if anyone here did. This story thus illustrates an important fact sometimes forgotten when reviewing these matters a quarter century later. At that time, an extremely thick cloak of secrecy surrounded all Soviet scientific activity. During the last Stalin years, no Soviet scientists (except for a few who

[4]The statement attacking Sakharov was in the form of a letter published in *Pravda*, August 29, 1973, p. 3.

went to special United Nations meetings) visited the West, and almost no Western scientists (other than an occasional member of the Communist party) visited the U.S.S.R. Not only was Soviet military-oriented research and development carried on in secret, but so also was all of what we would consider to be their unclassified work. Under such extreme circumstances, it was very easy for those inclined to believe the worst to do so. This situation began to change slowly following the death of Stalin, but by then the damage described in this book had been done.

Chapter FOUR

The Debate over the Superbomb

First Reactions

After the first Soviet atomic bomb exploded in Siberia on August 29, 1949, several days passed before the radioactive debris reached a point where it could be picked up by filters on U.S. aircraft which were making flights specifically for the purpose of finding such debris. It took additional time for the information to reach Washington. More days elapsed before a special committee of experts (Vannevar Bush, Robert Bacher, Robert Oppenheimer, and Admiral William Parsons) were able to decide whether a bomb had been exploded, or whether there had been some kind of reactor accident. And still more time went by while the experts explained their conclusions to the politicians. Thus it was that almost a month passed before President Truman informed Senator McMahon, chairman of the JCAE, about the Russian bomb and still one more day before the news was announced to the public on September 23.[1]

Most scientific experts had estimated that four to five years would be required for the Soviets to make a bomb, and, indeed, the time interval from the first U.S. test to the first Soviet test

[1]Truman's announcement and the Soviet response are reproduced in chapter 3.

was four years and six weeks. Even so, nearly everyone, including most governmental officials and most members of Congress, reacted to the event as if it were a great surprise. Most of them had either forgotten or had never known the experts' original estimates, and in any event the accomplishment simply did not fit the almost universal view of the U.S.S.R. as a technologically backward nation in which whatever creative spark existed was stifled by an oppressive state. Moreover, even in the minds of many of the expert estimators, the five-year period tended to move forward as time went on, and so even after four years had passed, they were thinking of the first Soviet bomb as still being several years off. And on top of being a great surprise, it was a singularly unpleasant one; it marked the end of a period in which the United States had enjoyed a monopoly on the most powerful weapon of all time, a monopoly which was seen by many as compensating for the difference between the hordes of conscripts available to the Communist bloc and the relatively smaller armies available to the Western countries then still in the process of trying to revive something in the style of their pleasant prewar way of life.

All of this raised a very serious pair of questions: What should the American response be, and how should we go about achieving it? There were two quite separate debates about these issues: one in public and one under an especially thick cloak of secrecy.

The public debate was relatively modest in scope and decorous in style, mainly due to the thick curtain of secrecy that surrounded the subject and the consequent public lack of detailed information. For all practical purposes, the public had never heard of the superbomb (which quickly became the main focus of the secret debate) when the argument began. The first public mention of this horrendous device came in a television speech by Senator Edwin Johnson carried locally in New York City on November 1, 1949. This was followed shortly after (November 18) by an article by Alfred Friendly in the *Washington Post* that used Johnson's speech as a starting point, and which gave some further details[1]. Later in January, when the secret debate was just about to reach its climax, an excellent article describing it was written by James Reston and published on page one of the *New York*

Times, but it came too late to influence the matter[2]. The public debate focused mainly on such questions as whether the Soviet explosion truly was a surprise and how serious it really was, and on such issues as the need for renewed attempts to achieve international control over atomic energy and the futility of excessive secretiveness as a means of preserving national security. As in the case of the 1946 debate over the organizational form of the postwar nuclear program, the *Bulletin of the Atomic Scientists* provided the main forum for the public debate.

In an article entitled "Needed: Less Witch Hunting and More Work," Harold Urey said,

> In spite of many statements on this subject over the past four years, apparently no progress has been made toward an understanding of the minor importance of that which Congress and public opinion regard as secrets. We scientists not only failed to convince Congress and the public of the soundness of our prediction that the Russians would have the bomb in about five years after we had it, but, because we told disagreeable truths, we have even been accused of wishing to give up our progress because we are impractical dreamers or plain traitors[3].

AEC Chairman David Lilienthal made a similar comment concerning secrecy:

> In part, the news means, to me, that we should stop this senseless business of choking ourselves by some of the extremes of secrecy in which we have been driven, extremes of secrecy that impede our own technical progress and our own defense[4].

Eugene Rabinowitch, founder and editor of the *Bulletin,* commented on both the need to keep ahead and on the need to negotiate:

> While we must do all we can to keep ahead in this race, we must continue looking for a large-scale imaginative political solution, which alone could stop the inexorable trend leading to atomic war. The conditions for renewed international control negotiations might not be better now than they were before, but they are sufficiently

> different to justify a complete review of the policy in this field, and an unprejudiced exploration of any new possibility which may offer itself[5].

And Linus Pauling, one of the strongest and most persistent voices among those fighting against the arms race, and at the time president of the American Chemical Society, said,

> I believe that the event should, however, serve to point out the necessity of taking immediate action to avert the atomic catastrophe that the world is facing. I believe that it should be a warning to the people of the world and a potent incentive to the nations of the world to resume negotiations, through the United Nations Organization for the establishment of an effective system of international control of atomic energy[6].

The secret debate about what the American response ought to be took place within the government itself. Many organizations were involved in it, including most importantly the National Security Council, the Departments of Defense and State, the JCAE, and the AEC. By virtue of the statutes establishing the last of these, it had the primary responsibility for generating U.S. nuclear policies and programs, and so most of the proposals and arguments about what to do arose within its precincts.

The early official reaction of the AEC's Los Alamos laboratory was a proposal to step up the pace of the nuclear weapons program across the board in all areas. This acceleration was to include tests designed to elucidate the possibilities of using small thermonuclear explosions as a means of boosting the efficiency of fission bombs as well as programs for investigating and exploiting more conventional means for further improving fission weapons. Among other measures, Norris Bradbury, the laboratory director, proposed that they go on a six-day work week, and that they expand the staff, especially in theoretical physics. The AEC's director of the Division of Military Applications, General James McCormack, received these proposals, and sought the GAC's advice concerning them.

Other AEC division heads were similarly studying proposals for expanding the relevant programs within their jurisdictions.

Laurence Hafstad, director of the Reactor Development Division, was reviewing plans for further expansion of plutonium production and for producing tritium for use in thermonuclear reactions. Walter J. Williams, director of production, generated proposals to expand production of U-235, and John K. Gustafson, director of raw materials, planned both to expand ore production from known sources in Africa and Colorado and also to initiate immediately a search for new sources.

At the same time, Teller, then at Los Alamos; Lawrence, Alvarez and Wendell Latimer at Berkeley; Robert LeBaron in the Pentagon, JCAE Chairman Senator Brian McMahon and his staff chief William L. Borden, and Lewis Strauss at the AEC had come to focus in on the super as the one correct answer to the Soviet A-bomb, and they initiated a concerted effort to bring the entire government around to their point of view as quickly as possible.

As discussed in chapter 2, the term "super" in those days referred to a proposal, dating from the early forties, for building a very large "superbomb" that would derive most of its energy from the reactions of deuterons—heavy hydrogen nuclei—with other deuterons or with tritons—a rarer, still heavier type of hydrogen nuclei not normally found in nature. Typically in early discussions of the super, explosive energy yields about one thousand times as great as those of "ordinary" atomic bombs were mentioned. Despite several years of thinking by some very bright people, no one then knew how to make a "super." About all that was then known for certain was that, in principle, the energy was there. However, it was generally, though not universally, believed that a concerted effort, parallel to the wartime Los Alamos program, would be crowned by success.

After a brief period of uncertainty the secret debate came to revolve about a single crucial issue: Was or was not a high priority program for the development of a superbomb the appropriate American response, and if so, how should we go about conducting such a program? Especially considering the enormity of the issue—and most of those involved were fully aware of its enormity—the participants in the secret debate were very few: the members of the GAC, the members of the AEC and a few of their staff, the members of the JCAE and a few of their staff, a very few top officials in the Defense Department, and a very small group of very concerned scientists, mostly from two of the AEC's laboratories. Altogether, there were less than one hundred people, many of whom thought of themselves—probably correctly—as being involved in making one of the most fateful decisions of all time.

The GAC's Conclusions

As a result of all this concern and churning about, the AEC called for a special meeting of its GAC to be held as soon as possible. The GAC was one of the special mechanisms established by the Atomic Energy Act of 1946 for the purpose of managing the postwar development of nuclear energy in the United States. Its function was to provide the AEC with scientific and technical advice concerning its programs. The members of the original committee were all persons who had been scientific or technological leaders in major wartime projects. Robert Oppenheimer, who was elected chairman of the committee, had been director of the Los Alamos laboratory where the first A-bomb had been designed and built. James B. Conant had been Bush's principal deputy on the Office of Scientific Research and Development (OSRD). Lee A. DuBridge had been the director of the radiation laboratory at M.I.T. Enrico Fermi, a Nobel Prize winning physicist, had been one of the earliest scientific workers in the field and had directed the construction of the first nuclear reactor at the University of Chicago in 1942. I. I. Rabi, another Nobel laureate, had been a leader at the M.I.T. Radiation Laboratory. Hartley Rowe had been a division director in the wartime National Defense Research Committee and a consultant at Los Alamos. Glenn T. Seaborg had been a chemist at the Metallurgical Laboratory. He was also codiscoverer of plutonium and later won the Nobel Prize in chemistry. He served longer than anyone else as chairman of the AEC (1961–1971); Cyril Stanley Smith had been a metallurgist at Los Alamos; and Hood Worthington of the DuPont Company, a leader in the construction of the Hanford reactors. Many of the members of this and later GAC's also served on other high-level advisory committees, including the Joint Research and Development Board of the Department of Defense (Conant, Oppenheimer, and Rowe) and the Scientific Advisory Committee to the White House Office of Defense Mobilization (Conant, DuBridge, Rabi, Oppenheimer). Some of them also served on some especially crucial *ad hoc* committees, and so a rather complex web of interlocking advisory committee membership soon developed. As a result, these individuals had much more influence than the simple sum of their memberships would indicate. In the fall of 1949, the membership of the GAC was still made up of the original group, with the exception of Oliver E. Buckley, president of the Bell Laboratories, who had replaced Hood Worthington.

Oppenheimer was not only the formal leader of the GAC, by virtue of his personality and background, he was its natural leader as well. His views were, therefore, of special importance in setting the tone and determining the contents of the GAC's reports in this as well as most other matters.

Throughout his service on the GAC he supported the various programs designed to produce and improve nuclear weapons. At the same time, he was deeply troubled by what he had wrought at Los Alamos, and he found the notion of bombs of unlimited power especially repugnant. Ever since the end of the war, therefore, he had devoted much of his attention to promoting international control over atomic energy with the ultimate objective of achieving nuclear disarmament. He and Rabi had been, in effect, the original inventors of the plan for nuclear arms control that later became known as the Baruch plan. Oppenheimer's inner feelings about nuclear weapons were clearly revealed in an oft quoted remark, "In some sort of crude sense which no vulgarity, no humor, no overstatement can quite extinguish, the physicists have known sin; and this is a knowledge which they cannot lose[7]."

The call for the meeting, in addition to raising the question of a high-priority program to develop the super, also asked the committee to consider priorities in the broadest sense, including "whether the Commission is now doing things we ought to do to serve the paramount objectives of the common defense and security[8]." The commission also asked for the GAC's views on plans for civil defense and the proposals for expanding production of plutonium and heavy water. And "as for the superweapon, the Commission wanted to know whether the nation would use such a weapon if it could be built and what its military worth would be in relation to fission weapons[9]." All members were present for the special meeting which was held on October 29–30, except Glenn Seaborg, who was in Europe. The GAC, however, had his views in the form of a letter.[2] The GAC in the course of its deliberations heard from a number of experts in various relevant fields, including George F. Kennan, counselor to the State Department and formerly U.S. ambassador to the U.S.S.R.; the chairman of Joint Chiefs of Staff; and nuclear experts Hans Bethe of Cornell University and Robert Serber of the University of California at Berkeley. Bethe had been the head of the Los Alamos Theoretical Division during the war, and was widely regarded as one of the most knowledgeable and wisest physicists in the field of nuclear energy and its applications. He was opposed to the development of the super as a response to the Soviet bomb. Serber was a distinguished physicist in his own right and had also been at Los Alamos, but in this particular instance he represented, in a limited sense, Lawrence and Alvarez.[3] Toward the end of the two-day meeting, the GAC had a long session with the commissioners themselves, and with their intelligence staff. The next day, the GAC prepared its report.

The GAC's report consisted of three separate sections plus two addenda. In 1974, the report was almost entirely declassified, with only a very few purely technical details remaining classified. The full report, minus such deletions, is presented in the Appendix.

Part I of the report dealt with all pertinent questions other than those directly involving the super. The GAC, in effect, reacted favorably to the proposals of the various AEC division directors with regard to expansion of the facilities for separating uranium isotopes and for the production of plutonium and for increasing

[2]In essence, Seaborg's letter says he would have to know a lot more about the matter before he could agree to oppose going ahead with the super (from the official report of the Oppenheimer hearings, pp. 238–239).

[3]Serber was always cool to the idea of building the super. He was at the GAC meeting principally to discuss a proposal of Lawrence and Alvarez' for building a special reactor for producing tritium (see chapter 7).

the supplies of ore. These proposals and the GAC's endorsement of them were followed eventually by a several-fold increase in the rate of production of fissile material.[4]

In Part I the GAC also recommended acceleration of research and development work on fission bombs, particularly for tactical purposes:

> TACTICAL DELIVERY. The General Advisory Committee recommends to the Commission an intensification of efforts to make atomic weapons available for tactical purposes, and to give attention to the problem of integration of bomb and carrier design in this field.

This quoted paragraph deserves special emphasis, since it is often suggested that Oppenheimer, Conant, and some of the others opposed nuclear weapons in general. They did apparently find them all repugnant, and they did try hard to create an international control organization that would ultimately lead to their universal abolition. However, in the absence of any international arms-limitation agreements with reliable control mechanisms, they explicitly recognized the need to possess nuclear weapons, especially for tactical and defensive purposes, and they regularly promoted programs designed to increase their variety, flexibility, efficiency, and numbers. For the next several years, right up to the time his security clearance was stripped from him, Oppenheimer continued strongly to promote the idea of an expanded arsenal of tactical nuclear weapons. The only type of nuclear weapon they opposed—and they did so openly—was the super. They also sometimes opposed certain other military applications of nuclear energy which they judged to be either grossly technologically premature (such as the nuclear airplane), or largely pointless (such as radiological warfare).

Part I of the report further recommended that a project be initiated for the purpose of producing "freely absorbable neutrons" to be used for production of U-233, tritium, and (perhaps) radio-

[4]The actual increase took place in two steps. The first step resulted solely from the stimulus of the Soviet A-bomb, and the second more-or-less-equal step came only after the Korean War reinforced that original impetus.

logical warfare agents. It recommended that the design of such a device be assigned to the Argonne National Laboratory, rather than to the Berkeley group which also wanted to do it (see chapter 7), and it added the further explanation that in the minds of its authors "[the] construction of neutron producing reactors is not intended as a step in the super program."

Perhaps most important of all in the present context, Part I also says, "We strongly favor, subject to favorable outcome of the 1951 Eniwetok tests, the booster program." This short phrase makes it abundantly clear that the GAC favored conducting research fundamental to understanding the thermonuclear process, and that its grave reservations, which we will take up next, were specifically and solely focused on one particular application of the fusion process, the enormously powerful and destructive superbomb.

Part II discussed the super. It outlined what was known about the super, and it expanded on the unusual difficulties its development presented, but it concluded it could probably be built. In part it said,

> It is notable that there appears to be no experimental approach short of actual test which will substantially add to our conviction that a given model will or will not work, and it is also notable that because of the unsymmetric and extremely unfamiliar conditions obtaining, some considerable doubt will surely remain as to the soundness of theoretical anticipation. Thus, we are faced with a development which cannot be carried to the point of conviction without the actual construction and demonstration of the essential elements of the weapon in question. This does not mean that further theoretical studies would be without avail. It does mean that they could not be decisive. A final point that needs to be stressed is that many tests may be required before a workable model has been evolved or before it has been established beyond reasonable doubt that no such model can be evolved. Although we are not able to give a specific probability rating for any given model, *we believe that an imaginative and concerted attack on the problem has a better than even chance of producing the weapon within five years.*

This last sentence (italics are mine) deserves special emphasis. It has been suggested in the past that the GAC in general and

Oppenheimer in particular were deceptive in their analysis of the technological prospects of the super, that is, that they deliberately painted a falsely gloomy picture of its possibilities in order to reinforce their basically ethical opposition to its development. Given the technological circumstances described in chapters 3 and 5, this statement of the program's prospects could hardly have been more positive.

The report then discussed what we might call the "strategic economics" of the super as then conceived:

> A second characteristic of the super bomb is that once the problem of initiation has been solved, there is no limit to the explosive power of the bomb itself except that imposed by requirements of delivery. . . . Taking into account the probable limitations of carriers likely to be available for the delivery of such a weapon, it has generally been estimated that the weapon would have an explosive effect some hundreds of times that of present fission bombs. This would correspond to a damage area of the order of hundreds of square miles, to thermal radiation effects extending over a comparable area, and to very grave contamination problems which can easily be made more acute, and may possibly be rendered less acute, by surrounding the deuterium with uranium or other material. It needs to be borne in mind that for delivery by ship, submarine or other such carrier, the limitations here outlined no longer apply and that the weapon is from a technical point of view without limitations with regard to the damage that it can inflict.
>
> It is clear that the use of this weapon would bring about the destruction of innumerable human lives; it is not a weapon which can be used exclusively for the destruction of material installations of military or semi-military purposes. Its use therefore carries much further than the atomic bomb itself the policy of exterminating civilian populations. . . . It is clearly impossible with the vagueness of design and the uncertainty as to performance as we have them at present to give anything like a cost estimate of the super. If one uses the strict criteria of damage area per dollar and if one accepts the limitations on air carrier capacity likely to obtain in the years immediately ahead, it appears uncertain to us whether the super will be cheaper or more expensive than the fission bombs.

In Part III they got to what to them was the heart of the matter, the question of whether the super should be developed:

Although the members of the Advisory Committee are not unanimous in their proposals as to what should be done with regard to the super bomb, there are certain elements of unanimity among us. We all hope that by one means or another, the development of these weapons can be avoided. We are all reluctant to see the United States take the initiative in precipitating this development. We are all agreed that it would be wrong at the present moment to commit ourselves to an all-out effort toward its development.

We are somewhat divided as to the nature of the commitment not to develop the weapon. The majority feel that this should be an unqualified commitment. Others feel that it should be made conditional on the response of the Soviet government to a proposal to renounce such development. The Committee recommends that enough be declassified about the super bomb so that a public statement of policy can be made at this time.

In the two addenda, those members of the committee who were present—that is, all except Seaborg—explained their reasons for their proposed "commitment not to develop the weapon." The first addendum was written by Conant and signed by Rowe, Smith, DuBridge, Buckley, and Oppenheimer. In part it said,

We base our recommendation on our belief that the extreme dangers to mankind inherent in the proposal wholly outweigh any military advantage that could come from this development. Let it be clearly realized that this is a super weapon; it is in a totally different category from an atomic bomb. The reason for developing such super bombs would be to have the capacity to devastate a vast area with a single bomb. Its use would involve a decision to slaughter a vast number of civilians. We are alarmed as to the possible global effects of the radioactivity generated by the explosion of a few super bombs of conceivable magnitude. If super bombs will work at all, there is no inherent limit in the destructive power that may be attained with them. Therefore, a super bomb might become a weapon of genocide.

The existence of such a weapon in our armory would have far-reaching effects on world opinion: reasonable people the world over would realize that the existence of a weapon of this type whose power of destruction is essentially unlimited represents a threat to the future of the human race which is intolerable. Thus

> we believe that the psychological effect of the weapon in our hands would be adverse to our interest.
>
> We believe a super bomb should never be produced. Mankind would be far better off not to have a demonstration of the feasibility of such a weapon until the present climate of world opinion changes.
>
> In determining not to proceed to develop the super bomb, we see a unique opportunity of providing by example some limitations on the totality of war and thus of limiting the fear and arousing the hopes of mankind.

Contrary to a frequently suggested notion, the members of the GAC were not at all unmindful of the possibility the Russians might develop the super no matter what the United States did. Indeed, they regarded it as entirely possible and explained why it would not be crucial:

> To the argument that the Russians may succeed in developing this weapon, we would reply that our undertaking it will not prove a deterrent to them. Should they use the weapon against us, reprisals by our large stock of atomic bombs would be comparably effective to the use of a super.

The minority addendum, signed by Enrico Fermi and I. I. Rabi expressed even greater revulsion to the super, but (weakly) coupled an American renunciation with a proposal for a worldwide pledge not to proceed:

> By its very nature it cannot be confined to a military objective but becomes a weapon which in practical effect is almost one of genocide.
>
> It is clear that the use of such a weapon cannot be justified on any ethical ground which gives a human being a certain individuality and dignity even if he happens to be a resident of an enemy country.
>
> The fact that no limits exist to the destructiveness of this weapon makes its very existence and the knowledge of its construction a danger to humanity as a whole. It is necessarily an evil thing considered in any light.
>
> For these reasons we believe it important for the President of the United States to tell the American public, and the world, that

> we think it wrong on fundamental ethical principles to initiate a program of development of such a weapon. At the same time it would be appropriate to invite the nations of the world to join us in a solemn pledge not to proceed in the development of construction of weapons of this category.

As in the case of the majority, Fermi and Rabi also explicitly took up the possibility the Soviets might proceed on their own, or even renege on a pledge not to:

> If such a pledge were accepted even without control machinery, it appears highly probable that an advanced state of development leading to a test by another power could be detected by available physical means. Furthermore, we have in our possession, in our stockpile of atomic bombs, the means for adequate "military" retaliation for the production or use of a "Super."

There is a minor contradiction in the report that merits special attention. The main body of the report says the minority feel that "the commitment not to develop . . . should be made conditional on the response of the Soviet Government . . .", whereas the minority addendum itself merely says ". . . it would be appropriate to invite the nations of the world to . . . join us in a solemn pledge not to proceed." This contradiction evidently arose from no more than a drafting problem. The report was drafted and edited in literally a matter of some hours, and it is remarkable for its overall clarity and consistency. In a recent conversation about this question, Rabi stated it was his firm recollection that he and Fermi definitely intended to couple American forbearance with a Soviet pledge to do the same. He said that he (and others on the GAC) saw the "super question" as providing an excellent opportunity to rekindle interest in the international control of all nuclear arms, not just the super alone. This recent recollection is completely consistent with Rabi's entire postwar record. He has served on many high-level advisory committees—including the Science Advisory Committee to the United Nations Secretary General—that have from time to time dealt with various elements of the overall nuclear arms control and disarmament problem, and he has always insisted on the need for much more progress in this area.

There were, of course, other places besides this report where various members of the GAC expressed their opinions about the matter. All such other views that I know of are consistent with those expressed in the report, and even where they may differ slightly, it would seem that their official and formal opinions as given in the report should be placed ahead of any more casual and individual statements.

Among such more casual statements is a famous letter from Oppenheimer to Conant, written shortly before the special GAC meeting and in which he said among other things:

> What concerns me is really not the technical problem. I am not sure the miserable thing will work, nor that it can be gotten to a target except by oxcart. It seems likely to me even further to worsen the unbalance of our war plans. What does worry me is that this thing appears to have caught the imagination, both of the Congressional and military people, as the answer to the problem posed by the Russians' advance. It would be folly to oppose the exploration of this weapon. We have always known it had to be done; and it does have to be done, though it appears to be singularly proof against any form of experimental approach. But that we become committed to it as the way to save the country and the peace appears to me full of dangers[10].

On December 2 and 3, five weeks after the special meeting, the GAC convened for one of its regularly scheduled meetings and carefully reviewed the question of the super once more. According to Richard Hewlett[11], Oppenheimer reported to the commissioners that no member wished to change the views expressed in the October 30 report. Four members and the committee's executive secretary (John Manley, of the Los Alamos staff) did, however, send separate additional papers to the commission expanding on their individual views. Hartley Rowe's paper argued that the dubious value of the super as a retaliatory weapon would not outweigh the dangers of diverting resources from other nuclear weapons development, helping the Russians in their development of a super, and undermining our nation's moral values. Fermi's and Manley's papers raised further questions about the possible military value of the super. Buckley reiterated his opposition to an immediate "all-out" effort on the super but he did call for a thorough and detailed study by the best scientists and mathematicians available on the design, delivery methods, and possible effects of the super. And DuBridge further challenged the military, psychological, and diplomatic value of the super.

In summary, and in my personal view, Oppenheimer and his colleagues based their opposition to the super on just two, for them, inseparable elements. First, even though they were accustomed to accepting the necessity for so-called nominal-sized atomic bombs, they felt that the super was simply too big and too murderous. Second, they believed its development was not necessary in order to insure American security. If either of these elements had been missing they would not have made the recommendation they did. If they had felt it was essential for national security they would have overcome (although probably with considerable regret) their ethical reservations. And if they had not been concerned about its excessive power they would not have so concerned themselves with the question of whether or not it was essential. Other factors certainly made it easier for them to reach their negative conclusion but were not essential to doing so. The most important such secondary factors were the competition for resources they felt were needed elsewhere in the nuclear weapons program, the great awkwardness and considerable uncertainty of such designs as did then exist, technically very promising alternative approaches to the design of bombs considerably larger than those then in the inventory but not so extremely large as the super, and the notion of using the special potential danger to mankind of the super as a basis for reopening nuclear arms control talks.

The Views of the AEC Commissioners

During the first part of November, the AEC Commissioners themselves met several times, including once with Oppenheimer and some others from the GAC, to discuss the super and the GAC's report on it, and to clarify their own views on the matter[12].

AEC Chairman David Lilienthal was receptive to the GAC's point of view. He similarly favored making two parallel responses to the Soviet test: one, increasing the production of fission weapons, and developing a greater variety of them particularly for use in tactical situations; and two, officially announcing our intention to refrain from proceeding with the super while simultaneously reopening and intensifying the search for international control of all kinds of weapons of mass destruction. Lilienthal considered the U.S. complete reliance on weapons of mass destruction as a fundamental policy weakness, and he viewed a crash program on the hydrogen bomb as foreclosing what might be the last good opportunity to base American foreign policy on "something better than a headlong rush into war with weapons of mass destruction." "We are," he said, "today relying on an asset that is readily depreciating for us, i.e., weapons of mass destruction. [A decision to go ahead with the super] would tend to confuse and, unwittingly, hide that fact and make it more difficult to find some other course[13]."

Lilienthal's views on this issue were fully consistent with his prior record. In 1946 he had chaired a committee whose purpose was to develop the details of the first U.S. proposals for the international control of nuclear energy. These proposals were accepted by Secretary of State Acheson and by the president with only a few modifications and became known as the "Baruch plan" when they were presented to the United Nations. Robert Oppenheimer was a member of that same committee and Lilienthal reports being very favorably impressed by his grasp of the situation. In 1947, after a long period of service as the first chairman of the Tennessee Valley Authority, Lilienthal became the first chairman of the AEC. During his tenure he frequently quarreled with Pentagon officials over the custody of nuclear weapons and over what he regarded as their unsupported requests for larger numbers of weapons. He also quarreled with the more conservative members of the JCAE over so-called management issues centering on secrecy and plant security.

Commissioner Henry DeWolf Smyth, the lone scientist member of the AEC, generally agreed with Lilienthal and the position of the GAC. He doubted the military value of super to the United States even if the Soviets did develop it. He agreed that the superbomb issue provided an excellent opportunity to reopen discussions of international control, and that such discussions would have a greater—though still small—chance of success if the United States announced in advance that it did not intend to develop the super. He did, however, reserve the right to reverse any decision to forgo the super within six months or a year.

Commissioner Sumner Pike, a former financier who had served in the Securities Exchange Commission and the Office of Price Administration, after some hesitation, fell in with Lilienthal and Smyth in opposing the development of the super at that time.

Commissioner Lewis Strauss strongly disagreed with the GAC's advice concerning the super. In addition, he felt that the committee had far overstepped the bounds of its competence in raising ethical issues, in evaluating the military usefulness of the super, and in discussing what the effects of its development would be on other peoples and nations. Moreover, during the course of the last several years, Strauss and Oppenheimer had openly as well as privately disagreed over a number of related issues and, I believe, this factor, which had a substantial personal dimension (Strauss found it extremely hard to admit a mistake), also played a role in Strauss' considerations.

In any event, Strauss himself had long been interested from a layman's point of view both in the applications of nuclear energy and also in military affairs, and he was the man most responsible for seeing to it that the aircraft and instruments that detected the first Soviet bomb were there to do so. He generally had confidence in what today is called "the technological fix" and so he readily concluded that the correct U.S. reply to the Soviet test was to make a "quantum jump" of its own, to seize once again the technological initiative, and he saw Edward Teller's proposal for a crash program on the superbomb as the ideal way to satisfy those goals. He regarded the proposal to forgo the development of the super as being tantamount to a proposal to choose deliberately to be a second-rate power. He expressed his views clearly and succinctly in a letter[14] to Truman, dated November 25, 1949:

> I believe that the United States must be completely armed as any possible enemy. From this, it follows that I believe it unwise to renounce, unilaterally, any weapon which an enemy can reasonably

be expected to possess. I recommend that the President direct the Atomic Energy Commission to proceed with the development of the thermonuclear bomb, at highest priority subject only to the judgment of the Department of Defense as to its value as a weapon, and of the advice of the Department of State as to the diplomatic consequences of its unilateral renunciation or its possession. In the event that you may be interested, my reasoning is appended in a memorandum.

In an addendum he gave the factual and technical bases as he saw them for his conclusions. Among these were,

- The production of such a weapon appears to be feasible (i.e., better than a 50-50 chance).
- Recent accomplishments by the Russians indicate that the production of a thermonuclear weapon is within their technical competence.
- A government of atheists is not likely to be dissuaded from producing the weapon on "moral" grounds.
- It is the historic policy of the United States not to have its forces less well armed than those of any other country (viz., the 5:5:3 naval ration, etc., etc.)

Commissioner Gordon Dean, while expressing his view neither so clearly nor so strongly, generally supported Strauss' position.

A report giving the views of the commissioners, was presented personally by Lilienthal to President Truman on November 9. The commission's report appended the full report of the GAC meeting of October 29-30, and also the individual views of the three commissioners in town that day (Lilienthal, Smyth, and Dean). Strauss, who was on business in California at the time, sent his views on to the president somewhat later in the form of the letter quoted just above.

Other Official Views

Senator Brian McMahon, the chairman of the JCAE, held views very similar to those of Lewis Strauss. McMahon, as chairman of the Special Senate Committee on Atomic Energy in 1945-1946, had been the leader in the development of the legislation (the Atomic Energy Act of 1946) that had led to the creation of the JCAE, the AEC, the GAC, and other such bodies. His subsequent career had been completely bound up with the elaboration of the U.S. nuclear program, and it was easy and natural for him to develop a proprietary attitude towards all nuclear matters and institutions. For some time, even before "Joe 1," McMahon had felt that we were not working as hard on nuclear weapons as the importance of strategic bombing warranted, and in July he had directed the committee staff, headed by William Borden, to catalogue all conceivable means for maximizing America's nuclear power. Borden had, as a result, concluded that what was called for was "a concerted effort to develop the ultimate weapon system—the thermonuclear weapon carried by a nuclear powered airplane[15]."

Prodded by Borden, a fanatic on the subject of nuclear weapons who four years later would formally accuse Oppenheimer of being a Soviet agent, McMahon frequently expressed his views in sometimes frantic and strident terms. The GAC report, he later told Teller "just makes me sick[16]." He noted that the super would produce more damage for less cost than fission weapons, and he said he could see "no moral dividing line between a big explosion which causes heavy damage and many smaller explosions causing equal or still greater damage[17]." He said that in the face of the Russians' great manpower the United States had no real choice. "If we let Russians get the super first, catastrophe becomes all but certain—whereas, if we get it first, there exists a chance of saving ourselves[18]." And "total power in the hands of total evil will equal destruction[19]."

At another point, McMahon said, "In my judgment, a failure to press ahead with the hydrogen bomb might well mean unconditional surrendering in advance—by the United States to alien forces of evil." Later in the same speech, he spoke more strongly about a hydrogen bomb program being needed in order to avert "well nigh certain catastrophe[20]."

This apocalyptic view of the matter was widespread. For example, Admiral Sidney Souers of the National Security Committee staff is reported to have said, "It's either we make it or we wait until the Russians drop one on us without warning[21]."

The chairman of the Research and Development Board of the Department of Defense, Karl T. Compton, also supported Lewis Strauss' position in a letter to President Truman. He did say that[22]

> If renunciation of this objective by the United States could ensure its abandonment or failure everywhere else in the world, I could agree with the recommendation of the GAC.

But he went on to say he saw no chance of that happening in the absence of a suitable means of monitoring Soviet work in the field, and concluded,

> Therefore, until an adequate international solution is worked out, it seems to me that our own national security and the protection of the type of civilization which we value, require us to proceed with the development of the most powerful atomic weapons which may be in sight. We can "hope to God they won't work," but so long as there is a reasonable possibility that they may work, it seems to me essential that we proceed with research and development on such projects as long as possible enemies may be doing the same thing.
>
> For whatever it is worth, therefore, my judgment is that we should proceed with this phase of atomic weapon development, with increased activity and support, but that we should do so without fanfare or publicity.

Also in the Pentagon, Phillip Morse, technical director of the newly formed Weapons Systems Evaluation Group, studied the question and reached similar conclusions which he reported to Compton and the Joint Chiefs of Staff.

Robert LeBaron, chairman of the Military Liaison Committee[5] and special assistant to the Secretary of Defense for Atomic Energy, was in close personal touch with the pro-superbomb

[5]The Military Liaison Committee was another of the special mechanisms established by the Atomic Energy Act of 1946 for the purpose of managing the American nuclear program. It was located in the Pentagon, its chairman (usually a civilian) was regarded as a Pentagon official, and its other members were general and flag officers from each of the services. Its main duty was to ensure that the separately organized and civilian-managed AEC's nuclear weapons programs were adequate for national security.

scientists, and became the Pentagon's strongest proponent of a crash program on the hydrogen bomb.

The Joint Chiefs of Staff also favored going ahead with the super, but they were not as enthusiastic about it as many of its civilian supporters were. They were also slower in formulating their ideas in the matter, but they finally did conclude that if the Soviets were to come into sole possession of such a weapon, the position of the United States would be intolerable[23]. They therefore urged determination of its technical feasibility, but they saw no need for a crash program to do so. They opposed forswearing the super, and as for the moral issues raised by the GAC, they said it was folly to argue in war that one weapon was more moral than another.

Defense Secretary Louis Johnson accepted the views of Compton, Morse, LeBaron, and the Joint Chiefs of Staff, who together were his top civilian and military advisors in such matters, and he, too, joined forces with Strauss in promoting the H-bomb.

The Pro-Superbomb Scientists

The views of those politicians favoring the H-bomb program—Strauss, McMahon, LeBaron, and Johnson—were based in large part on the lobbying being done by three nuclear scientists: Edward Teller, Ernest Lawrence and Luis Alvarez. Those three were not, of course, the only scientists favoring a crash program on the super, but of those who did, they had by far the best access to political figures. For instance, John Von Neumann also lobbied for the bomb, but he was then not as well connected as those named and so his efforts consisted mainly of "talking [the] ear off" Robert Oppenheimer and, perhaps, other highly placed science advisors like him[24].

For Teller, of course, promoting the H-bomb was nothing new. In 1942 and 1943, even before the Los Alamos laboratory was created, Teller had been very interested in such a device, and he continued his interest after the Manhattan project was fully under way. It was the area he intended to emphasize most strongly when he briefly considered staying on at Los Alamos as head of the Theoretical Division after the war. In 1947, he referred cryptically to the super in the *Bulletin of the Atomic Scientist:*

> Actually, it is quite unsound to limit our attention to atomic bombs of the present type. These bombs are the results of our first attempts and they were developed under wartime pressure. The paramount consideration had to be: which of the developments promised earliest results. In a subject as new as atomic power we must be prepared for startling developments. It has been repeatedly stated that future bombs may easily surpass those used in the last war by a factor of a thousand. I share this belief[25].

He continued to promote the idea whenever he could with AEC officials and others. By chance, he was in Washington at the Pentagon on the day Truman announced the Soviet test, and he called Oppenheimer to ask him what he should do to help meet the new Soviet challenge. According to Teller, Oppenheimer simply told him to "Keep your shirt on." As a result, Teller struck off on his own, and from that day on he promoted the idea of a crash program with all of those he contacted: Oppenheimer and the other scientists, General McCormack and other officers and officials in the AEC and the Pentagon, and perhaps most importantly, members of the JCAE. His message was simple: If a superbomb can be built, and it probably can, it can be built just as readily by the Russians as by us; if we do not mount a high-priority program to build one, then the Russians may very well succeed in doing so first, and that would result in certain disaster for us. So persistent was Teller, that Stanislaw Ulam, who was very close to many of the principals on both sides of the debate, is said to have commented that "some of the opposition to the super might have been a reaction against Teller's insistent advocacy of the new weapon[26]."

Supporting Teller in his political campaign to win support for the super were Ernest Lawrence and Luis Alvarez.[6] Lawrence

[6]Alvarez, fully realizing that he had suddenly become involved in an extraordinary matter, kept a diary of his activities. He later used extensive quotations from his diary in his testimony in the Oppenheimer hearings. See Alvarez' testimony, USAEC, *In the Matter of J. Robert Oppenheimer*, The M.I.T. Press, 1971, pp. 770–805.

was one of America's most honored experimental scientists, and director of the University of California's Radiation Laboratory (UCRL). His laboratory had been one of the major elements in the wartime Manhattan project, and he himself had been a member of the highest scientific advisory councils during the war. After the war, his laboratory prospered as a basic research institution, and he continued to play a major role in influencing the course of the United States nuclear programs, but he did so largely through close but unofficial personal contacts in the JCAE, the Pentagon, and the AEC rather than through membership on formal advisory bodies. Lawrence was what today would be called a "technological optimist." He also believed that the sole business of scientists was to produce new knowledge and technology, and that what was done with it afterwards was entirely the responsibility of politicians. Even in the case of the debate on the super, he did not think of himself as a political proponent of the super; rather he saw himself as simply opposing those scientists who were trying to stifle work on it for their own political, and therefore improper, reasons. When Oppenheimer made his famous remark about how "the physicists have known sin; and this is a knowledge which they cannot lose," Lawrence replied, "I am a physicist and I have no knowledge to lose in which physics has caused me to know sin[27]." Alvarez was a protégé of Lawrence's and one of the leading physicists in Lawrence's laboratory. During the war, he had worked both on radar at M.I.T. and the A-bomb at Los Alamos. Shortly after the war he invented a radar-controlled aircraft landing system called GCA (Ground Controlled Approach) which played a crucial role in breaking the 1948 Berlin blockade. As a result of that work he, too, had developed some important Washington connections. On hearing the news about the Soviet test, Lawrence, Alvarez, and a third colleague at Berkeley, Wendell Latimer, became very concerned both about what the overall American reaction ought to be, and also about what the response of their own laboratory should be. They quickly decided that Teller's proposal for a concerted attack on the superbomb was the appropriate national response. They made several trips to Los Alamos, mainly to discuss the matter with Teller, and to Washington to promote the idea and to explore the question of what they themselves could and should do. In the course of doing so they managed to exploit their government connections very effectively. As a result, Lilienthal in his diary referred to the pro-H-bomb arguments as the "E. O. Lawrence-Strauss line:[7] If we don't get this super first, we are sunk, the U.S. would surrender without a struggle[28]," and at another point he refers to "E. O. Lawrence and Luis Alvarez in here drooling over the [super][29]." Alvarez in his diary put it very differently: "October 5, 1949: Latimer and I independently thought the Russians could be working hard on the super and might get there ahead of us. The only thing to do seems to be to get there first—but hope that it will turn out to be impossible[30]."

The permanent members of the Los Alamos staff and the regular visitors to the laboratory also generally reacted negatively to the GAC report. John Von Neumann, a frequent visitor whose opinions always carried very great weight, favored a high-priority program for the development of the super. Carson Mark, the head of the Theoretical Division, did not advocate accelerating the work on the H-bomb, but he resented what he took to be being told in effect by the GAC that his division should suspend the studies it had been making concerning the super ever since the end of the war. Most others at Los Alamos seem to have held views similar to those of one or the other of these leaders. As Teller later put it, "The GAC report seemed to [say] . . . : As long as you people work very hard and diligently to make a better bomb, you are doing a fine job; but if you succeed in making real progress toward another kind of nuclear explosion, you are doing something immoral. To this, the scientists at Los Alamos reacted psychologically. They got mad, and their attention was turned toward the thermonuclear bomb, not away from it."

Truman's Decision

Well aware of the serious conflict building up over this issue within the government, Truman turned to a special subcommittee of the National Security Council (NSC) for assistance. This sub-

[7]It is interesting to note that Lilienthal at no point mentions Teller. As far as Lilienthal was concerned, Lawrence was the scientific focus of the pro-H-bomb forces.

committee consisted of AEC Chairman Lilienthal, Secretary of Defense Johnson, and Secretary of State Dean Acheson. It had been established in early 1949, well before Joe 1, in order to do a better job of correlating military requirements with AEC development and production than had been done hitherto. The first meeting of the subcommittee's working group (that is, of staff representing the principals) on the question of the super took place on November 30, and the first meeting of the subcommittee itself took place a few weeks later.

In the subcommittee's considerations of the matter, Lilienthal continued to support the conclusion of the GAC and the 3–2 majority of the commissioners (himself, Pike, and Smyth versus Strauss and Dean) and to urge restraint while making one more attempt to achieve an international agreement to forgo the development of the super. He said he felt that the "whole purpose and course of mankind was tied to this decision[31]."

Louis Johnson duly supported the views of his civilian and military advisors and urged that the United States proceed as rapidly as possible with development of the H-bomb. His views of the world and the hydrogen bomb's role in it can be clearly seen in some public comments he made four days after Truman reached his decision in the matter. Referring to the U.S.S.R., he said, "There is but one nation in the world tonight that would start a war that would engulf the world and bring the United States into war . . . We want a military establishment sufficient to deter that aggressor and sufficient to kick the hell out of her if she doesn't stay deterred[32]."

Acheson sided with Secretary Johnson. He had been one of those responsible for developing the Baruch plan for controlling nuclear energy at the end of the war, and he had participated in various ways in the fruitless international debate over the matter ever since. As a result of his experiences as a "cold warrior," he was deeply pessimistic about the possibility of achieving any useful agreements with Stalin and the Soviets in the matter of the super. He also supported the view of his staff that sole possession of the super would severely damage not only our military position but our foreign policy position as well. In addition, by January 1950 he came to believe that the pressures building up for a firm decision (really, for a decision favoring the super program) had reached a point where it could not be delayed any longer[33]. Acheson's understanding of and opinion about the main arguments opposing the development of super are given in his book *Present at the Creation*[34].

> Enough evil had been brought into human life, it was argued by men of the highest standing in science, education, and government, through development of atomic weapons without adding the super-horror of thermonuclear ones. If the United States with its vast resources proved that such an explosion was possible, others would be bound to press on to find the way for themselves. If no one knew that a way existed, research would be less stimulated. Those who shared this view were, I believed, not so much moved by the power of its logic (which I was never able to perceive—neither the maintenance of ignorance nor the reliance on perpetual goodwill seemed to me a tenable policy) as by an immense distaste for what one of them, the purity of whose motive could not be doubted, described as "the whole rotten business."

Not all State Department officials held views like Acheson's. In particular, George Kennan, a very distinguished and influential diplomat and scholar who specialized in Soviet and Eastern European affairs, took a position similar to that of Oppenheimer. In his memoirs[35] he says, "a number of us, including the late Robert Oppenheimer, felt that before proceeding with the development of weapons of a wholly new range of destructiveness we should reexamine our situation with respect to the international control of atomic weapons generally, and make sure that there was really no possibility of arriving at international agreements that would obviate the need to embark upon this fateful course." On (or about) January 20, shortly after he had relinquished his part as chairman of the State Department Policy Planning staff, he submitted a personal memorandum[8] supporting such views to Secretary Acheson.

---

[8]As he later recalled it, his memo to Acheson including the recommendation that "we remain prepared to go very far, to show considerable confidence in others, and to accept a certain risk for ourselves, in order to achieve international agreement on their [that is, nuclear weapons'] removal from international arsenals."

The final meeting of the Special Committee of the NSC was held on January 31, 1950. Secretary Acheson presented the other members with a draft set of recommendations for the President[36]:

> (a) That the President direct the Atomic Energy Commission to proceed to determine the technical feasibility of a thermonuclear weapon, the scale and rate of effort to be determined jointly by the Atomic Energy Commission and the Department of Defense: and that the necessary ornance developments and carrier program be undertaken concurrently;
>
> (b) That the President defer decision pending the reexamination referred to in (c) as to whether thermonuclear weapons should be produced beyond the number required for a test of feasibility;
>
> (c) That the President direct the Secretary of State and the Secretary of Defense to undertake a reexamination of our objectives in peace and war and of the effect of these objectives on our strategic plans in the light of the probable fission bomb capability and possible thermonuclear bomb capability of the Soviet Union;
>
> (d) That the President indicate publicly the intention of the Government to continue work to determine the feasibility of thermonuclear weapon, and that no further official information on it be made public without the approval of the President.

At the urging of Secretary Johnson, the second paragraph was stricken. Lilienthal, apparently realizing that stopping the super project had become impossible, accepted the recommendations as being the best compromise that could be achieved at the time. However, immediately after presenting these recommendations to the president, in the afternoon of that same day, Lilienthal requested and was given the opportunity to once again state the "grave reservations I had to the course recommended." He told the president that he did not expect his opinion to be of much value in the "face of agreement by the Secretary of State and the Secretary of Defense," but that he could not avoid saying he believed "this course was not the wisest one and that another course was open[37]." (Lilienthal's influence in the matter may also have been undermined by the fact that he had already submitted his resignation from the AEC some months before, and he would be leaving the post in only a matter of days. His resignation was not, it should be emphasized, directly connected with the superbomb question.)

In the meantime, on January 27, Klaus Fuchs, then working in England, but formerly one of the members of the British team at Los Alamos, and one of the participants in the spring 1946 conference on the super, confessed that he had engaged in espionage on behalf of the U.S.S.R. between 1942 and 1949. The GAC held a special meeting on January 30 to assess what Fuchs knew and what he could have passed on. They concluded that it could have been a great deal, and this fact was passed on to the special subcommittee of the NSC for consideration at its meeting on January 31.

According to Richard Hewlett, the records of the meetings immediately before the president's final decision do not show that the knowledge of Fuchs' treachery had an influence on the thinking of the principals or the conclusions they reached[38]. Similarly, Lilienthal's rather lengthy diary entry covering these meetings never mentions Fuchs. And according to the later recollections of Gordon Arneson[39], who attended the January 31 NSC meeting, "the Fuchs matter was in the back of everyone's minds, but not dominant," when the committee met. On the other hand, James R. Shepley and Clay Blair, Jr., claimed that the treachery of Fuchs "dominated and stampeded" the discussion[40]. In any event, the position of the three principals—Acheson and Johnson for moving ahead, Lilienthal for restraint—did not change as a result.

Later on that same day, January 31, 1950, President Truman announced his decision to go ahead with the development of the H-bomb.

> It is part of my responsibility as Commander in Chief of the Armed Forces to see to it that our country is able to defend itself against any possible aggressor. Accordingly I have directed the Atomic Energy Commission to continue its work on all forms of atomic weapons, including the so-called hydrogen or superbomb. Like all other work in the field of atomic weapons, it is being and will be carried forward on a basis consistent with the over-all objectives of our program for peace and security.

> This we shall continue to do until a satisfactory plan for international control of atomic energy is achieved. We shall also continue to examine all those factors that affect our program for peace and this country's security[41].

Even though the words simply instructed the AEC to "continue its work on all forms . . . including the . . . superbomb," those who had opposed the super took them to mean defeat, and those who favored an accelerated program to develop the super (with the possible exception of Teller) generally took them to mean victory for their point of view. In any event, whatever ambiguity may have been contained in Truman's January 31 decision was removed only a few weeks later. On February 24, 1950, the Joint Chiefs of Staff requested the president to approve "all out development of hydrogen bombs and means for their production and delivery[42]." Truman again asked the Special Committee of the NSC for its advice. (Sumner Pike had by then replaced Lilienthal.) A week later, the Special Committee agreed with the Joint Chiefs of Staff and recommended "that preparations be made for the quantity production of the H-bomb without waiting for the results of the test" then tentatively planned (see the program discussion in chapter 5). On March 10 the president so ordered, and the construction of the reactors for producing the tritium thought to be necessary commenced soon after.

At the same time, and consistent with the last paragraph in the announcement, the president also set in motion a study "to reexamine the national objectives in peace and war, and the effect on these aims of the new Soviet nuclear capabilities demonstrated by the detonation of August 1949." This directive led to the well-known rearmament study, NSC 68[43].

The decision was widely acclaimed in Congress and the press. As the *New York Times* put it, "No Presidential announcement since Mr. Truman entered the White House seemed, in the opinion of many observers, to strike such an instant or general chord of nonpartisan congressional support.

"The repeatedly expressed theme was that regardless of how dreadful the hydrogen weapon might be, Mr. Truman had no other course in view of the failure so far of negotiations for international control of atomic energy and of the 'atomic explosion' some months ago in the Soviet Union[44]." Only very few members of Congress expressed any reservations; one particularly notable case being W. Sterling Cole, a Congressman from upstate New York and a member of the JCAE. The *New York Times* reported that he "asserted that the President had usurped Congressional authority and had acted against the recommendations of competent authorities. 'The security, not simply of the United States but of mankind is at stake,' he said[45]."

The responses of the knowledgeable scientists were mixed. Some, as was to be expected, endorsed the decision and went on to emphasize that the answer to the question of whether or not a hydrogen bomb could be built depends on the facts of nature, and like it or not, one about which we have no choice. Thus, Edward Teller in an article entitled "Back to the Laboratories" said[46]:

> The scientist is not responsible for the laws of nature. It is his job to find out how these laws operate. It is the scientist's job to find the ways in which these laws can serve the human will. However, it is *not* the scientist's job to determine whether a hydrogen bomb should be constructed, whether it should be used, or how it should be used. This responsibility rests with the American people and with their chosen representatives.

Harold Urey sounded the same note[57]:

> It is unnecessary to emphasize the unpleasantness of such weapons. Many would wish that such weapon developments as these should prove physically impossible; but nature does not always behave in the way we desire. I believe we should assume that the bomb can be built.
>
> I am very unhappy to conclude that the hydrogen bomb should be developed and built. I do not think we should intentionally lose the armaments race; to do this will be to lose our liberties, and, with Patrick Henry, I value my liberties more than I do my life.

The Federation of American Scientists again warned about the new dangerous turn the arms race was taking and said,[9]

> We repeat now our request that the President establish without delay a new commission, with the broad perspective of the Acheson-Lilienthal Commission of 1946, to examine the whole issue of our atomic policy and to make a fresh start, looking toward a policy which offers some real hope of breaking the present stubborn deadlock.

A group of the leading U.S. physicists,[10] meeting in New York, signed a call for a "no first use" pledge:

> We urge that the United States, through its elected government, make a solemn declaration that we shall never use this bomb first. The circumstance which might force us to use it would be if we or our allies were attacked by this bomb. There can be only one justification for our development of the hydrogen bomb, and that is to prevent its use.

Others expressed their objections to the great secrecy in which the decision was made. Oppenheimer, the central figure in these supersecret deliberations, said,

> There is grave danger for us in that these decisions have been taken on the basis of facts held secret. This is not because the men who must contribute to the decisions, or must make them, are lacking in wisdom; it is because wisdom itself cannot flourish, nor even truth be determined, without the give and take of debate

[9]Clifford Grobstein, who was executive secretary of FAS at the time, also reports that he and some others from that organization met with Oppenheimer in December 1949, at the time of the Westinghouse Awards banquet, to discuss the whole situation with him. They knew then that the trend of events was leading towards a decision to develop the H-bomb, and they urged Oppenheimer to resign in protest to try and prevent it. As Grobstein recalls, Oppenheimer's reply was to the effect that he could do more to help bring about the results they all agreed were desirable if he continued to remain on the "inside."

[10]The signers were S. K. Allison, K. T. Bainbridge, H. A. Bethe, R. B. Brode, C. C. Lauritsen, F. W. Loomis, G. B. Pegram, B. Rossi, F. Seitz, M. A. Tuve, V. F. Weisskopf, and M. G. White. Most played leading roles in the development of U.S. weapons systems including the A-bomb during World War II.

> or criticism. The relevant facts could be of little help to an enemy; yet they are indispensable for an understanding of questions of policy[48].

Similar concern about the dangers of making such decisions by a limited group of men under such secretive conditions were voiced by Arthur Compton in the *Bulletin*, and by Louis Ridenour, Hans Bethe, and Robert Bacher in the *Scientific American*[49].

In the AEC and among its facilities, the decision and the deliberations leading up to and immediately following it brought about a number of major, far-reaching consequences. Los Alamos did accelerate both its H-bomb program and its program to improve fission bombs, the AEC production of special nuclear materials and nuclear bombs was very substantially increased, the Berkeley group instituted several programs designed to augment the Los Alamos program, and the programs designed to achieve nuclear-driven ships and airplanes were accelerated.

Curiously, perhaps, in the Department of Defense the decision had very little direct effect. The Defense budget for fiscal year 1951, which was constructed within the executive branch during the very same months it was struggling with the H-bomb question, and which was debated in Congress during the several months following the president's decision, was completely unaffected by either the Soviet A-bomb or the U.S. reaction to it. It took the sudden onset of the Korean War some four months after Truman's H-bomb decision before the Defense Department arms acceleration started.

However, some important studies of what it all meant were conducted. One of these was directed by Paul Nitze, chairman of the State Department Policy Planning Group, and ultimately resulted in a report known as NSC 68,[11] basically a call for a

[11] Nitze has been influential in national security matters throughout most of the quarter century following NSC 68. He served variously as asst. sec. of defense, sec. of the navy, and deputy sec. of defense under Kennedy and Johnson, and as a member of the SALT negotiations team under Nixon. NSC 68's origins and results have been the subject of an exceptional amount of study. The reasons probably include the accessibility of some of the principals to scholars and its apparent influence on subsequent policy.

very substantial and broadly based rearmament program. That call at first fell on deaf ears, only receiving a positive response after the Korean War started. Following that event, something very similar to the kind of rearmament called for in NSC 68 would, of course, have happened anyway.

Truman's decision and the series of events before and after it also affected the JCAE and its role in the governance of the U.S. program. The JCAE, and its staff, correctly saw themselves as having played a key and active role in the preparation of the executive decision. For decades after the decision, members of the joint committee on occasion portrayed their role in the matter as being one in which they saved the country from a most serious situation by overruling the advice of Oppenheimer and the GAC[50]. During the secret debate over the super, the committee dealt with AEC laboratory officials and with many other AEC scientists from within the ranks so to speak; that is with scientists like Teller and Alvarez who had no official responsibility but who were well informed on certain important issues and had strong opinions on the subject. This practice has continued ever since, and is part of the procedural pattern that has made it possible for the JCAE to get so deeply and effectively into the programmatic details of the AEC program. Several of the long-term members of the committee became quite expert at certain elements of the program—for example, Senator Henry M. Jackson on weapons and reactors, Representative Melvin Price on the nuclear airplane, Senator Clinton Anderson on whatever Los Alamos and the Sandia laboratory were doing—and have been strong and effective advocates of certain programs and projects. The net result is that, for good or ill, the AEC's weapons programs have always progressed much more rapidly than they otherwise would have.

Chapter FIVE

The Superbomb Programs

The American Program

Within a year after Truman announced his decision to "continue working on the Superbomb," the number of physicists in the Los Alamos T-Division (Theoretical Division) had increased to thirty-eight. Looked at one way, that was double the number that had worked there before the Soviet explosion, scarcely a simple continuation of the prior work. Looked at another way, it was only a few percent of all the theoretical physicists in the United States, hardly the mobilization that some had felt was needed.

The quality of the new staff was high. Among those who joined the regular staff were Frederick DeHoffmann, Frank Hoyt, Burton Freeman, Conrad Longmire and Marshall Rosenbluth. Among those who worked at Los Alamos for at least a full year in this period of 1949–50 were George Gamow, Lothar Nordheim, Edward Teller, and John Wheeler; while Hans Bethe, Enrico Fermi, Richard Garwin, Emil Konopinski, and John Von Neumann all consulted for at least several months in this period.

In the cases of Fermi and Bethe, this may seem surprising, since each had actively opposed going ahead with the hydrogen bomb before the president decided the issue. Those who know

Fermi say he apparently felt that once legitimate political authorities decided the question, it was no longer appropriate for the scientist to interject his own personal opinion. And in Bethe's case, he himself made it clear he still hoped that construction of an H-bomb would prove to be impossible, and he wished to help prove that to be the case. Also, of course, the sudden onset of the Korean War occurred only four months after Truman's decision and helped to further tip the balance in doubtful cases.

In addition, two small groups of theorists employed at other research centers were doing directly relevant but largely unclassified work in support of the Los Alamos program. One such group was located at the Argonne National Laboratory of the University of Chicago and worked between 1946 and 1952 on problems associated with the properties of materials at extremely high temperatures. Maria Mayer, who was also at Chicago, was closely associated with this group. The other group was at Yale University and was headed by Gregory Breit. It worked on certain basic interactions of particles and radiation.

Even with the theoretical group thus greatly augmented, Los Alamos still found itself unable to come up with a good idea as to how to meet the basic goal of the super program: ignite a relatively large (in principle, arbitrarily large) mass of thermonuclear fuel using a relatively small quantity of fission fuel. There were, however, a number of ideas—arising in part out of the Booster studies—concerning how to accomplish the reverse, that is, how to ignite a relatively small mass of thermonuclear fuel using a relatively large mass of fissile fuel.

As mentioned earlier, in 1948 a full-scale test of a device incorporating the booster principle was put on the list of shots to be included in the Pacific test operation already scheduled for 1951. On the basis of work undertaken largely before the first Soviet explosion, it became possible in January 1950 to choose a specific model for this experiment and to begin calculations designed to work out the final details of the design and to predict the details of the sequence of events that were expected to take place during the explosion. An experimental explosion of a device incorporating the booster principle and designated as the "Item" shot was carried out at Eniwetok on May 24, 1951. It was successful and led to much further elaboration of the idea, but it is outside the scope of this work (and beyond the limits imposed by classification as well) to here pursue further the discussion of this particular line of development.

In addition to the booster experiment, in late 1949 consideration was begun of a closely related but somewhat different device in which a relatively large fission yield was to be used to ignite a relatively small mass of thermonuclear fuel in a pattern better adapted for the observation of phenomena expected to be associated with the super. These studies, which were very actively pursued from that time on, resulted in the design of the device tested in an experiment known as the "George" shot of the 1951 Greenhouse series. Reduced to its essentials, the purpose of this experiment was to show, as a minimum, that a thermonuclear reaction could under ideal conditions be made to take place in an experimental device. This experiment came to play a key role in the super program. As Teller later put it ". . . we needed a significant test. Without such a test no one of us could have had the confidence to proceed further along speculations, inventions, and the difficult choice of the most promising possibility. This test was to play the role of a pilot plant in our development[1]."

GEORGE AND ITS SIDE EFFECTS

The test that Teller was referring to, the George shot, took place on May 8, 1951. It worked. The largest fission explosion to date succeeded in igniting the first small thermonuclear flame ever to burn on earth.

The year preceding the test had been one of intense work among both theoretical and experimental physicists, some of them members of the regular Los Alamos staff, some of them consultants resident there for a few months at a time, and still others working elsewhere as subcontractors to the Los Alamos laboratory. The theorists and mathematicians worked mainly on the problem of the basic design, that is, how to configure the device, how to get sufficient energy from the exploding mass of fissile material into the thermonuclear fuel in order to achieve the temperature necessary to make it burn. The experimental physicists and engineers worked mainly on two other problems; how to actually build the configuration the theorists called for, and how to make certain measurements, called diagnostic measurements, whose purpose was to observe in detail the various processes as they unfolded. Several outside groups were engaged to assist in making these diagnostic measurements; two of these were at the Naval Research Laboratory and one was at the UCRL at Berkeley.[1] Working under the overall direction of Frederick Reines of the Los Alamos staff, these three groups, plus several others based at Los Alamos itself, observed the phenomena that took place in the first fraction of a microsecond after the explosion was initiated. Even though the contribution of the thermonuclear reactions to the total energy released was relatively small, these experiments confirmed that the device had functioned according to plan.

In the meantime, some months before the George shot took place, some of the theorists finished their immediate, George-related tasks and found time to think again about the next step: the ignition of a large mass of thermonuclear fuel by a relatively small fission explosion. Many ideas were tossed out; some absurd, some clever and interesting but impractical, and a few close to the mark. As Teller later put it in an article entitled "The Work of Many People"[2]:

> Two signs of hope came within a few weeks; one sign was an imaginative suggestion by Ulam; the other sign was a fine calculation by DeHoffmann.
>
> I cannot refrain from mentioning one particularly human detail in DeHoffmann's work. Since I had made the suggestion that led to his calculation, I expected that we would jointly sign the report containing the results. Freddie, however, had other plans. He signed the report with my name only and argued that the suggestion counted for everything and the execution for nothing. I still feel ashamed that I consented.

---

[1]Hugh Bradner and I, both experimental physicists at UCRL, Berkeley, directed a diagnostic measurement program which was developed at Berkeley and executed at Eniwetok Atoll. This program later became one of the main elements leading to the formation of the Lawrence Livermore Laboratory, the second American nuclear weapons laboratory founded in 1952.

Ulam put it somewhat differently. Referring to the same period[3], he recalls,

> . . . I thought of a . . . scheme, and after I put my thoughts in order and made a semi-concrete sketch, I went to Carson Mark to discuss it. Mark, who was head of the theoretical division, was already in charge of the very extensive theoretical work supporting on a very large scale Teller's and Wheeler's special groups. The same afternoon I went to see Norris Bradbury and mentioned such schemes. He quickly grasped its possibilities and at once showed great interest in having it pursued. Next morning, I spoke to Teller.
>
> At once Edward took up my suggestions. Hesitantly at first, but enthusiastically after a few hours. He had seen not only the novel elements, but had found a parallel version, an alternative to what I had said, perhaps more convenient and general. From then on pessimism gave way to hope. In the following days I saw Edward several times. We discussed the problem for half an hour or so each time. I wrote a first sketch of the proposal. Teller made some changes and additions, and we wrote a joint report very quickly. It contained the first engineering sketches of the new possibilities of starting thermonuclear explosions. We wrote about two parallel schemes based on these principles. The report became the fundamental base for the design of the first successful thermonuclear reactions and the test called "Mike." . . . A more detailed follow through report was written by Teller and DeHoffmann.

These two descriptions are different, but they are not really contradictory. The "suggestion [I made] that led to his calculation" in the Teller article is precisely the "parallel version, an alternative . . . perhaps more convenient and general" of Ulam's recollection.

The details of this idea, which are still highly classified, are set down in two reports. One is dated March 9, 1951, and signed by Ulam and Teller[4]. The other report, issued less than a month later, included some extensive calculations by Frederick DeHoffmann, but as Teller says above, is signed by Teller only[5]. This new idea[2] was immediately recognized by all concerned as very

[2]It should be clear that this "new idea" is *not* the idea of using lithium deuteride in place of liquid deuterium. That idea was already several years old at the time. Rather, it concerns how to configure the device so that a relatively small fission explosion can ignite an arbitrarily large amount of thermonuclear fuel. There has been so much confusion on this point that I feel it is useful to emphasize it again here.

promising.[3] Calculations based on it proceeded forthwith. Most of them were done by Los Alamos scientists, using whatever methods and computing machines they could lay their hands on. In addition, the physics department at Rand Corporation (Richard Latter and others), and, most importantly, a newly assembled group at Princeton, assisted in these calculations. The Princeton group, assembled and directed by John Wheeler,[4] was known as Project Matterhorn-B[5] (B for bomb). It made some especially significant contributions, using some very new computing machines available in nearby cities. One was an early model of the UNIVAC still in the final check-out phase at the factory of Sperry-Rand at Philadelphia. Another was the MANIAC, a pioneer computer developed by Von Neumann. The latter was located nearby in Princeton, and was one of the rationalizations behind setting up the group there. However, even though it pioneered many of the most important ideas involved in modern digital computing, it was too late in coming into full operating condition to contribute much to the super program[6].

In June 1951, the new design idea and the calculations supporting it were presented to a wider group in a meeting held in Oppenheimer's office at the Institute for Advanced Study at Princeton. In attendance were members of the AEC and of the GAC, plus Teller, Bradbury, Bethe, and other staff and consultants of the Los Alamos laboratory. It was again immediately recognized by this wider audience that this new idea was the way to go. Several years later, Oppenheimer compared the ideas presented in June 1951 with those that had been current earlier in the

---

[3]Teller described this idea to me in late April or early May during a visit to Eniwetok where I was engaged in setting up one of the diagnostic experiments for the George test. I recall that it was immediately evident to me that it would work, that here was the long-sought solution to the problem.

[4]The group was made up of some very bright young theorists including Richard Bellman, John Toll, Lawrence Willets, Edward Freeman, Louis Henyey, and Army Capt. Carl Hausmann, who later became one of the leaders at the Lawrence Livermore Laboratory. Robert Oppenheimer, John Von Neumann, and Eugene Wigner all worked in the vicinity, but only rarely dropped in.

[5]At about the same time another research group, called Matterhorn-S was also formed. The S stood for Stellerator or (Lyman) Spitzer, and that group worked on one particular approach to the production of thermonuclear energy in a magnetically confined plasma.

following words[7].

> The program we had in 1949 was a tortured thing that you could well argue did not make a great deal of technical sense. It was therefore possible to argue also that you did not want it even if you could have it. The program in 1951 was technically so sweet that you could not argue about that. The issues became purely the military, the political and the humane problem(s) of what you were going to do about it once you had it.

Shortly after, plans were developed at Los Alamos and approved in Washington for a series of tests of devices designed to verify and exploit these ideas. The program for the construction and test of what was to be the first very large thermonuclear device was placed in the hands of Marshall Holloway and other long time staff members of the laboratory.

In the fall of 1951, Teller left Los Alamos and returned to his faculty post at the University of Chicago. He left in part because the main problem, from the point of view of a theorist at least, had been solved, but, more importantly, he left because of growing animosities and arguments between himself and the laboratory administration over how and by whom the thermonuclear program should be run. Also, leaving the laboratory made it easier for him to promote his own political views in Washington about what should be done. Thus, he worked very closely with the air force and certain elements in the AEC on the promotion of a second nuclear weapons laboratory to be set up in order to provide an alternative to and competition with the Los Alamos laboratory.

At about that same time, it was decided that there should be two separate test operations. One, IVY, was to be held as soon as possible, and had as its main objective verifying the Teller–Ulam configuration in a strictly experimental situation. The other, CASTLE, would come as soon after IVY as the technological circumstances permitted, and had as its main objective verifying the Teller–Ulam configuration in forms suitable for regular military use.

As the AEC's director of the Division of Military Applications put it in a now declassified official planning document[8]:

Operations IVY and CASTLE are each planned to include the firing of a complete experimental thermonculear device. IVY is principally directed toward establishing thermonuclear feasibility from a purely experimental device and providing important fundamental information. CASTLE, however, is further directed toward a deliverable emergency weapon in addition to furnishing important information.

In the event IVY is successful, it is anticipated that it will become a matter of great urgency to provide the deliverable version of a thermonuclear weapon with minimum delay. Although CASTLE is tentatively scheduled to occur within one year of IVY, there is no certainty that the lag between an experimental device and a deliverable weapon, which are physically quite different, can be shortened to this extent. The importance of this time lag is such as to suggest it may be a determinant for peace or war.

The short term goal in the thermonuclear field should be to create a deliverable thermonuclear weapon with a minimum lag between experimental device and usable weapon. This means gambling on the success of [one of the CASTLE devices] as we are doing. . . . It will, on the one hand, provide back-up for an emergency design and, on the other hand, speed the ultimate design(s).

. . . The thermonuclear program has as a short term objective the successful demonstration of the first deliverable thermonuclear weapon in order to provide an emergency military capability in this field. The long term objective will be the development of optimum thermonuclear weapons, which will proceed along the lines of reducing weapon size and weight, and improving efficiency and utility, as has taken place in the field of fission weapons.

MIKE

The first experimental test of a large thermonuclear device took place on November 1, 1952, as one of the two tests conducted during Operation IVY. It was very successful. The Mike shot, as the explosion itself was called, yielded an energy equivalent to that contained in 10 million tons, or 10 megatons, of TNT, an amount roughly 1000 times as large as that released by the Hiroshima bomb (some 13,000 tons or 13 kilotons), and thus remarkably close to the predictions made for the size of hydrogen bombs since the beginning of the study of them in the forties. The Mike device was, however, very far from being a practical deliverable weapon. Its thermonuclear fuel consisted of liquid deuterium, a normally gaseous substance which must be cooled to temperatures colder than −250°C in order to liquify it. It required, therefore, a very complex refrigerating device in order to maintain it in that condition prior to its being exploded. The entire mechanism was very large and bulky, occupying a substantial laboratory building especially constructed to house it on the island of Elugelab in Eniwetok Atoll.

There were several reasons for using the difficult—from an engineering point of view—liquid deuterium as the thermonuclear fuel in the first major test rather than the very much more convenient saltlike solid lithium deuteride (LiD). One reason was that the most influential persons in determining how the bomb should be built were theoretical physicists, and from the point of view of physical theory, it was much easier and more straightforward to understand and calculate the simple "burning" of deuterium than the complex multi-step, multi-branched chain-like process involved in the explosion of LiD. In a pioneering experiment involving a really new idea, such as Mike truly was, that kind of simplification can be very helpful. Another reason involved the fact that the naturally occurring form of lithium is not the optimal form for use in a thermonuclear explosion. Natural lithium contains only 7.4 percent Li^6, and it is only the Li^6 that produces tritium upon reacting with neutrons and thus contributes to sustaining the overall reaction process. It is cheap enough to produce lithium enriched in this isotope in very large quantities once a plant for doing so is under way, but it takes time and a very considerable capital investment in order to get production started. For these reasons and others, it was decided to conduct the first test with straight deuterium and delay the verification of the usefulness of LiD in the Teller–Ulam style of device until the following test series, a little more than a year later.

KING

A second device (the King shot) tested at Operation IVY on November 16 is also of great interest, especially from a political point of view. It was dropped from an aircraft and exploded at an altitude of 1480 feet above its target. It produced an extremely large explosion, but it derived its energy entirely from fission. Its actual yield has never been officially released, but 500 kilotons is surely a very good estimate. According to testimony given by David Lilienthal during the Oppenheimer hearings, the idea of testing a 500 kiloton device during Operation IVY was being considered even before he left the commission in early 1950. According to President Dwight D. Eisenhower in a speech made a year after IVY the United States then possessed an atomic bomb 25 times the size of the ones used in Japan, meaning 500 kilotons. And the height to which the King shot's mushroom cloud rose is consistent with a 500 kiloton explosion.

The political basis for developing and testing such a large fission bomb lay in the sometimes strident claims by some American politicians and physicists after "Joe 1" to the effect that the only possible way for America to ensure its security was to produce a much bigger bomb than the Russians had, and that the right way to do that was through the crash development of the super. Hans Bethe, in opposing the crash program for the super, argued that even if a much bigger bomb were the correct political answer, such a bomb could readily be produced by a straightforward extrapolation of the then current fission technology, and hence a crash program for the development of the superbomb was not needed for that purpose. (There is a special bit of irony in this matter. Although the King-type bomb was originally suggested as an alternative to the super, as has so often happened in the technological arms race, both programs were pursued and both were successful, and the number of different ways of achieving mass destruction was once again needlessly increased.)

The technical basis for Bethe's views lay in the exploratory research work that the Los Alamos laboratory had done on fission bombs since the end of the war. One part of this work was of an experimental nature and included laboratory research designed to improve implosion techniques, as well as the series of experimental test explosions conducted during Operation Sandstone in 1948. The other, theoretical part of the work involved the use of mathematical calculations to explore the possibilities of new configurations. As time went on during the late forties, both new computational techniques were invented and more powerful computing machines became available, with the result that such calculations not only could be done faster but also produced more reliable results. Among other things, these calculations clearly indicated that very efficient fission bombs in the megaton class could be made by adopting the latest implosion techniques for that purpose. One of the key persons in promoting and performing such calculations was Theodore B. Taylor. His particular contributions later (in 1974) received much publicity and attention in connection with quite a different nuclear issue prominent at the time[9].

The IVY King test confirmed these ideas by producing an explosion equivalent in its energy release to a half million tons, or half a megaton, of TNT.[6] As will be seen below, this particular test assumes very special importance when a comparison of the American and Soviet programs for the development of very large nuclear weapons is made.

BRAVO

The second test series mentioned in the AEC planning document was known as Operation CASTLE, and took place in the spring of 1954. No less than six variants of the superbomb were tested. The very first test in that series, the Bravo test, was of a device using LiD as its fuel and yielding 15 megatons. It was in a form readily adaptable for delivery by aircraft, and thus was the first large American hydrogen bomb. Several other devices tested in that series were also in forms suitable for military use, and yet another air-deliverable bomb, based directly on the Mike device,[7] was in a standby status to be deployed as needed if all others failed.

Table 1 gives the explosive yields and other pertinent data concerning the six tests that made up the CASTLE series.

From that point on, the development of the other lighter and more efficient weapons called for in the AEC planning document

[6] In a profile of T. B. Taylor in the *New Yorker* magazine by John McPhee (December 2, 1973), a device of this sort is referred to as the "Super Oralloy" bomb. Oralloy means Oak Ridge Alloy, or U-235.

[7] In addition to Marshall Holloway and other Los Alamos staff, Richard Garwin also played a key role in working out the design modifications needed to convert Mike into a practical weapon.

TABLE 1 Nuclear Tests at Operation CASTLE

1. Bravo	Land	1 March	14.8	Megatons
2. Romeo	Barge	27 March	11.0	Megatons
3. Koon	Land	7 April	0.1	Megatons
4. Union	Barge	26 April	6.9	Megatons
5. Yankee	Barge	5 May	13.5	Megatons
6. Nectar	Barge	14 May	1.69	Megatons

Adopted from Joint Force Seven, *Report WT-933: Cloud Photography*, January 27, 1958. (Originally classified as Secret, Restricted Data; declassified 12-31-72, and now in the Los Alamos Report Library.) The yields given here (except Koon) are derived from a single phenomenon, namely cloud measurements. Except for Bravo and Koon, the yields given in the table are not available in any other secondary sources to my knowledge. The Koon yield is from the 1962 edition of Glasstone's *The Effects of Nuclear Weapons;* the small yield and the existing natural cloud formation did not permit an estimate of its yield by Cloud Photography. Koon is the Livermore "fizzle" referred to elsewhere in this book, and in Teller's *The Legacy of Hiroshima.*

proceeded rapidly, and versions suitable for a variety of missiles quickly followed.

It is important to note in this connection that the air force decision to conduct a crash program to build an intercontinental missile—the Atlas—to deliver a hydrogen bomb was made in early February 1954, several weeks *before* the CASTLE test series began. It was made, therefore, on the basis of the IVY results and the pre-CASTLE calculations, and not on the basis of the CASTLE results. I emphasize this point and also the large number and variety of the CASTLE tests because they provide further confirmation of the fact that it was the Mike explosion that proved the basic idea, and that Bravo and other CASTLE tests were more in the nature of confirmations of the applicability of the Mike design in a number of other variants having different dimensions and using different more practical fuels.

Since those beginnings the Teller–Ulam idea has been much further elaborated, and a great variety of bombs incorporating both fission and fusion reactions have been tested and produced. In general, in the United States these developments have been directed towards the development of much lighter and physically smaller bombs having somewhat less explosive power for use in such modern devices as the so-called MIRV (*M*ultiple *I*ndependently-targetable *R*e-entry *V*ehicle) warheads for the Minuteman and Poseidon missiles. As far as I am aware, the United States has never exploded a device bigger than Bravo. In fact, when one such more powerful device was proposed in the mid-fifties by the Livermore laboratory for inclusion in a test series, General A. D. Starbird (then director of the Military Applications Division of the AEC) told me that Eisenhower personally vetoed it on the grounds of its excessive power.

The Soviet Program

According to Igor Golovin, Kurchatov's longtime colleague, deputy, and biographer, the success of the first Soviet A-bomb test in 1949 also inspired the Soviet government to institute a major, high-priority program to develop a hydrogen bomb[10].

> After a short celebration [following the August 29 explosion] Kurchatov did not at once send the participants in the test back to their homes. He held them all over for a week to give them time to calm down, and to prepare for the second stage in the atomic epoch which was still standing before them.
>
> In the American press, frightening stories were appearing about a super bomb allegedly being prepared in the U.S.A. Our government asked Kurchatov and other physicists to explain the basis of this blackmail. There was a basis. In addition to the uranium and plutonium bombs based on the fission chain reaction, another kind was possible—a hydrogen bomb, in which hydrogen is converted to helium, and which released much more energy for an equal weight of charge.
>
> So far, not all the strengths of Soviet physicists had been mobilized in the creation of the atomic bomb. Kurchatov recommended bringing in scientists of the old generation who had so far been standing aside from atomic affairs, and new young scientists and engineers whom the war had overtaken in universities and institutes. Up to the day of the test of the first atomic bomb, the work of the theorists had proceeded steadily, and showed that the hydrogen weapon

> might be a hundred or a thousand times more powerful than the atomic.
>
> After two months of rest, Kurchatov began the new attack. Sometimes in Moscow and sometimes distant from there, he helped speed the development of atomic industry. He put in brief periods back in his laboratory No. 2.[8] Here he gave first priority to that work which was most important for atomic industry and which was necessary for the creation of the hydrogen bomb.

The earliest of the "frightening stories" referred to by Golovin appeared in the *Washington Post* on November 18, 1949, less than three months after the Russian A-bomb and therefore just about the time "Kurchatov began the new attack[11]." It is possible—even probable—that the early American stories and Truman's subsequent decision to go ahead with a program in the United States influenced the scale and priority of the Russian program, but it is most unlikely they were its root cause. The Russians, like us, had evidently been seriously pondering the possibilities for some years and they too fully realized that a successful high efficiency fission explosion was a necessary precondition for the development of a thermonuclear bomb. Furthermore, and probably more importantly, they had known through Klaus Fuchs in 1946 (or before) that American scientists had been seriously studying this matter since 1942.

The only other direct statement concerning the timing of the Soviet thermonuclear weapon program is one made by Andrei Sakharov in 1974[12]. Sakharov has been widely heralded as the "Father of the Soviet H-Bomb." He himself says he was indeed "the author or coauthor of several key ideas," but that the *sobriquet* "very inaccurately reflects" his role. He also informs us that "a few months after defending my dissertation in the Spring of 1948, I was included in a research group working on the problem of a thermonuclear weapon." Standard Soviet biographical sources say that Sakharov was elected to the prestigious Soviet Academy of Science in 1950 (at age twenty-nine) for his contributions to the program to produce thermonuclear energy under controlled

[8] The "name" of the Moscow laboratory. Laboratory No. 1, a less important installation, was in Kharkov.

conditions. That could be true, but election at such an exceptional age would seem more plausible if it were based, at least in part, on contributions to the Soviet weapons program.

These data would seem to indicate that Sakharov and at least a few others began working on the theory of thermonuclear weapons somewhat ahead of the time Kurchatov "began the new attack." There is not, however, any necessary contradiction between Golovin's and Sakharov's remarks. Indeed, it seems to me that Sakharov's earliest work probably corresponds in some way to the theoretical studies that had been going on at Los Alamos ever since 1943, and that the "new attack" Golovin refers to corresponds roughly either to the substantial increase in theoretical work in the summer of 1949 or to the big increase in all kinds of effort that followed Truman's January 1950 decision.

AUGUST 1953—JOE 4

On August 8, 1953, Soviet Premier Georgy Malenkov announced that the United States not only had no monopoly on the A-bomb, it had no monopoly on the H-bomb, either. Just four days later in the early morning of August 12, 1953, the first test of a Soviet thermonuclear device took place. This was the fourth Soviet nuclear explosion whose occurrence was announced by the United States, and it is therefore commonly known as "Joe 4." As in the case of the U.S. tests, the Soviet scientists made experimental observations of the detailed processes of the explosion (presumably much like the U.S. diagnostic experiments) in order to understand what happened "in case the explosion turned out to be weak."

The explosion took place on a tower, the purpose, evidently, being to reduce the fallout hazard which would be created by the "tens of kilograms of radioactive products that are formed in the explosion." The test was observed by "the scientists, designers and workers who prepared the bomb" and by "administrators and representatives of the Party and the Government" and "Military Supreme Command." When it exploded "the heart of the Chief Theorist[9] beat wildly." The explosion vaporized the steel

[9] Not otherwise identified here, but from other accounts surely Andrei Sakharov.

tower and its concrete footings, and left a "huge crater" in its place. A number of scientists, including Kurchatov, were able to drive directly into the crater in tanks soon after the explosion, so it was evidently more like the shallow bowl-like depression often seen following U.S. tower shots rather than the deep steep-walled craters produced even by relatively small—a few kilotons—bombs placed directly on or under the surface. After the explosion, the area surrounding the point where the tower had been was covered by "yellow lumpy glass" which became thinner further from the epicenter.

No official figure for the yield of this first Soviet thermonuclear explosion has ever been given by anyone. The official U.S. compilation of Soviet explosions which occasionally does present rough estimates[13], simply says in this case that it was "thermonuclear." Soviet Premier Malenkov in confirming U.S. reports of the test, said it showed that " . . . the power of the hydrogen bomb is many times greater than the power of the atomic bomb[14]." Since Soviet A-bombs in those days had yields of some 20 kilotons, the use of the word "many" presumably sets a lower limit for the yield of this explosion of around 100 kilotons. Malenkov's announcement is not of equal help in estimating an upper limit, but it should be noted that the word "many" would be a peculiarly soft and modest adjective to use in boasting about the event if the device had yielded anything even remotely like Mike whose "power" was a 1000 times "greater than the power of the atomic bomb."

A rough value in the upper limit can, however, be estimated from information that came out long after the explosion. In this connection, we first note that the Soviets did not claim to have bombs yielding a megaton or more until after their next series of thermonuclear tests in 1955, and then Khrushchev boasted about it as an "important new achievement." In addition, the description of the explosion quoted above also can be used to estimate a rough upper limit of its size. The U.S. Bravo explosion of 1954 yielded 15 megatons of explosion energy and produced roughly four hundred kilograms of radioactive fission products plus a comparable amount of other radioactive materials from tritium to the transuranic elements, plus many others in between. Thus, by simply scaling, the "tens of kilograms" of fallout expected from Joe 4 means they expected a (presumably modest) multiple of 170 kilotons. Similarly, the description of the crater also implies an explosion of a hundred or some hundreds of kilotons, rather than megatons. If this explosion took place on a tower no more than 100 meters in height, then the explosion could not have been much more than roughly 400 kilotons if it left a depression which had a melted surface, and which was shallow enough and firm enough to be navigated by a tank. If the explosion had been much bigger, a relatively steep-sided crater with a surface consisting of loose rubble would have been produced instead. In basing this estimate of an upper limit on the use of a 100 meter tower, I am, I think, being fairly conservative, since this is somewhat taller than the tallest towers (300 feet) the United States found practical for such heavy devices in those days. If the tower was in fact shorter than 300 feet, then the upper limit would be correspondingly lower.[10]

Putting all this information together leads to an estimate that the explosion yield of the first Soviet thermonuclear explosion, Joe 4, was several hundred kilotons, probably though not certainly somewhat smaller than King, the largest U.S. all-fission bomb, and less than one-twentieth the size of Mike, the U.S. thermonuclear explosion of the previous year. This is admittedly, not an ideal way to make estimates of the size of a nuclear explosion, but until someone who knows how big Joe 4 was will say, it is the best that can be done. Furthermore, I believe the uncertainty in this kind of estimate, which is, I suppose, a factor of two or so, is small enough to allow sensible deductions about the political or military significance of the explosion.

As in the case of the explosive yields, no official statement has ever been made concerning the particular thermonuclear fuel used in this first Soviet thermonuclear device. However, here again it is easy to deduce that it very probably was lithium deuteride.

[10]According to Glasstone, ed., *The Effects of Nuclear Weapons*, Washington, D.C.: USAEC, 1962, p. 39, a megaton or more on a 450-foot tower will blast out a crater of the second type. Using a cube root scaling law, which ought to be reasonable for this modest range of variables, leads to the estimate of an upper limit of 400 kilotons for a 100-meter tower.

Whenever the Soviets feel moved to make a comparison between their 1953 Joe 4 experiment and the U.S. 1952 Mike shot, they always point out that although Mike came earlier, it was not a "real hydrogen weapon" because it was "only a very cumbersome and untransportable structure based on principles unsuitable for producing a weapon." The only "principle" involved in Mike that possibly might make it "unsuitable" as a weapon was its use of liquid deuterium instead of lithium deuteride as its thermonuclear fuel. By exclusion then, it is obvious that Joe 4 used lithium deuteride.

The second Soviet thermonuclear test took place on November 23, 1955. It was dropped from an aircraft and exploded high above the ground. The official U.S. statement about its yield is that it was in the "megaton range[15]." Fortunately for our understanding of both this device and the earlier Joe 4, Soviet Party Chairman Khrushchev was in India on an official visit at the time. He could not help boasting about it. He said,[16]

> I shall not say there has not been such an explosion. It was a terrific explosion . . . .
>
> These tests have demonstrated important new achievements by our scientists and engineers. The latest explosion of an H-bomb has been the most powerful to this time. Using a relatively small quantity of fissionable material, our scientists and engineers have managed to produce an explosion equal to several megatons of conventional explosives.

I do not, of course, know the precise configuration of this particular Soviet bomb. However, the words "important new achievement" which made it possible by "using a relatively small quantity of fissionable material . . . to produce . . . several megatons" make it as certain as it can be in such circumstances that this was the first application by the Russians of the Teller–Ulam idea or something much like it. Even so, the word "several" suggests it might have had only, say, one third the yield of Mike, the first U.S. application of these same ideas.

Since those early experiments, the Soviets also have produced further elaborations on this same theme. However, it took them several more years following their 1955 test before they were able to match the explosions in our 1952 and 1954 tests. Later, after they suddenly put an end to the nuclear test moratorium of 1958–1961, they tested some still larger bombs, one of which is said to have yielded 58 megatons, though neither we nor they have claimed that such a bomb is currently stockpiled.

Chapter SIX

The General Advisory Committee's Advice in Retrospect

A Comparison

In the fall of 1952, about the time of the U.S. Mike explosion, the JCAE solicited the views of a number of scientists on the question of how far ahead the United States was in thermonuclear developments. The answers ranged from zero to six years[1]. Since the Soviet program and the American program clearly followed different paths, no single answer of "x years" can be given even in retrospect, but it is now evident that the people whose answers were in the middle of this range were closest to the truth.

The following chronological comparison of the American and Soviet H-bomb programs is based on the foregoing program descriptions, and provides the simplest answer that can be given even now to the joint committee's question.

1. Both countries initiated high priority programs for the development of an H-bomb at about the same time—late 1949/early 1950—and both had been seriously studying the subject for some years before that.
2. The first U.S. tests designed to verify that a small thermonuclear reaction could be ignited by a large fission explosion took place

at Eniwetok, May 8 and May 24, 1951. Plans for inclusion of a device with this general objective were laid in 1948, before the first Soviet A-bomb. The final designs of the devices actually used for this purpose were worked out in 1950. They were successful.

3. The first test of a device configured to ignite a large thermonuclear explosion by means of a relatively small quantity of fissile material took place at Eniwetok, November 1, 1952. The device was known as the Mike test, it did succeed, and it produced a tremendous explosion equivalent in its energy release to 10 million tons (or 10 megatons) of TNT, or as had been repeatedly predicted since the early forties, roughly "1000 times larger" than the first atomic bombs. For certain practical reasons relating to the pioneering nature of the test, this first version of the Teller–Ulam configuration used liquid deuterium as its thermonuclear fuel.
4. Also in November 1952, the United States tested a very powerful fission bomb, code-named "King," and yielding one half of a megaton of explosive energy. Its purpose was to provide the United States with an extraordinarily powerful bomb by making a straightforward extension of fission weapon technology should such large bombs become necessary for any strategic or political reason.
5. The first Soviet explosion involving thermonuclear reactions took place on August 12, 1953. It was only some tens of times as big as the standard atomic bombs of the day, about the same size as but probably smaller than King, the largest U.S. fission bomb. It evidently involved one of several possible straightforward configurations for igniting relatively small amounts of thermonuclear material (as compared to the U.S. Mike and Bravo devices) with a relatively large amount of fissile material. It was the first device anywhere to use lithium deuteride as a fuel, and presumably could readily have been converted to a practical weapon if there were any point in doing so. It seems to have been a development step bypassed by the United States in its successful search for a configuration which made it possible (in Khrushchev's 1955 words) to produce a large explosion with a relatively small quantity of fissionable material.
6. In the spring of 1954, the United States successfully exploded six variants of the superbomb in Operation CASTLE. Their yields varied from a low of about one tenth of a megaton up to 15 megatons (see Table 1). The first and most famous of these tests was code-named Bravo, and it was exploded on March 1, 1954, at Bikini Atoll. Its design, which was initiated before the Mike explosion,

also incorporated the Teller–Ulam configuration, but used the more practical lithium deuteride as its thermonuclear fuel. Its yield was 15 megatons, even bigger than Mike, and it was readily adaptable for delivery by aircraft.

7. On November 23, 1955, the Soviets exploded a bomb having a few megatons of yield, and involving an "important new achievement" which made it possible "[by] using a relatively small quantity of fissionable material . . . to produce an explosion of several megatons." This device was evidently the first one in which the Soviets incorporated the Teller–Ulam configuration or something like it. It used lithium deuteride as a fuel. It was therefore a true superbomb, comparable to the U.S. Bravo device exploded 20 months earlier, except for its yield, which was still probably only about one-fifth the yield of Bravo.

The above chronology is summarized in chart form in the top time-line in Figure 1. Note that all of the U.S. explosions from George to Nectar, except Item, are the American reaction (or, better, overreaction) to the actual Joe 1 and the speculations about what would follow it. As a measure of the degree of overreaction, the number and power of this group may be compared with the actual 1953 and 1955 Soviet explosions charted. Note that except for King, only explosions involving fusion reactions are shown. If fission explosions were included, the overall balance would remain essentially unchanged.

Some Conjectures

With this chronology in mind, what can one say about what might have happened if the United States had followed the advice of Robert Oppenheimer and the GAC, and David Lilienthal and the majority of the AEC commissioners, and had not initiated a program for the specific purpose of developing the super in the spring of 1952?

At best, the invention of very large relatively inexpensive bombs of the super type would have been forestalled or, at least, substantially delayed. Very probably, the work on the booster principle, which we must assume would still have gone forward, would have led eventually to the ideas underlying the design of very big bombs, but these ideas might well have been delayed until both Eisenhower and Khrushchev were in power. These two leaders were both more seriously interested in arms limitation agreements than their predecessors had been, and it is at least possible that they might have been able to cope successfully with the superbomb. To be sure, such a favorable result was not very probable, certainly it had much less than an even chance of coming about, but its achievement would have been so beneficial to all mankind that it was clearly worth striving for and it was worth at least running some small risk.[1]

To evaluate just how much risk was involved we next examine the most probable alternative world and the "worst plausible" alternative world. (These two alternatives, along with the actual world are presented in Figure 1.)

In both of these alternative worlds I assume that the United States would have forgone the development of the super, but that the Soviets would have ignored American restraint, and would have proceeded at first just as they in fact did in the actual world. I also assume that the United States would have vigorously followed the positive elements of the GAC's advice, that is, the booster project and other ideas for improving fission bombs would have been accelerated. The difference between the most probable alternative world and the worst plausible alternative lies in the timing of the test of the first Soviet superbomb. In the worst plausible world I assume this test would have come on the same date that it did in the real world. In the most probable alternative world, however, I assume that this test would have been substantially delayed. The reasoning behind these two assumptions will be explained below.

In both of these two alternative worlds, then, in August 1953 the Soviets would have exploded "Joe 4," a large bomb deriving part of its explosive energy from a thermonuclear fuel and yielding

[1]Indeed, after several false starts which included Eisenhower's "Atoms for Peace" and "Open Skies" proposals, as well as certain Soviet initiatives in 1955–1957, Eisenhower and Khrushchev did achieve the nuclear test moratorium of 1958–1961, and this moratorium in turn was the political ancestor of all the arms control agreements achieved since that time.

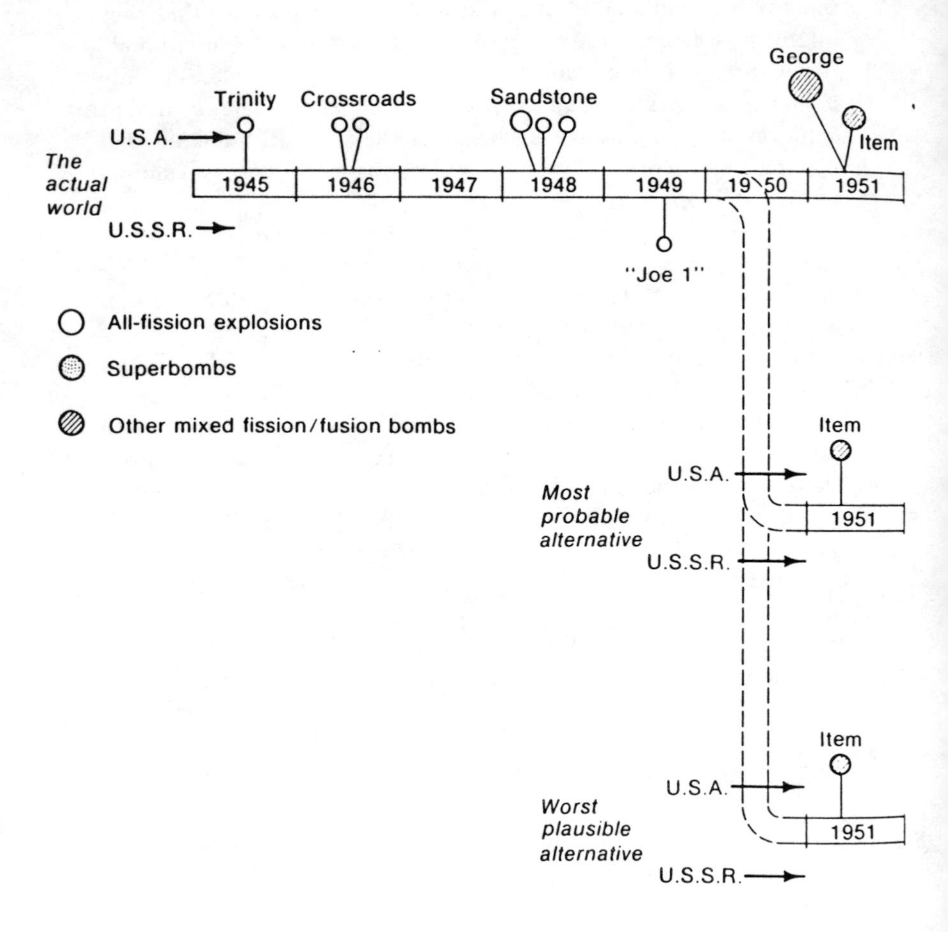

FIGURE 1
Chronology of U.S. and Soviet H-Bomb Program. (Areas of circles are proportional to the areas that would be destroyed by the bombs indicated. After 1950 all "nominal" size tests are omitted.)

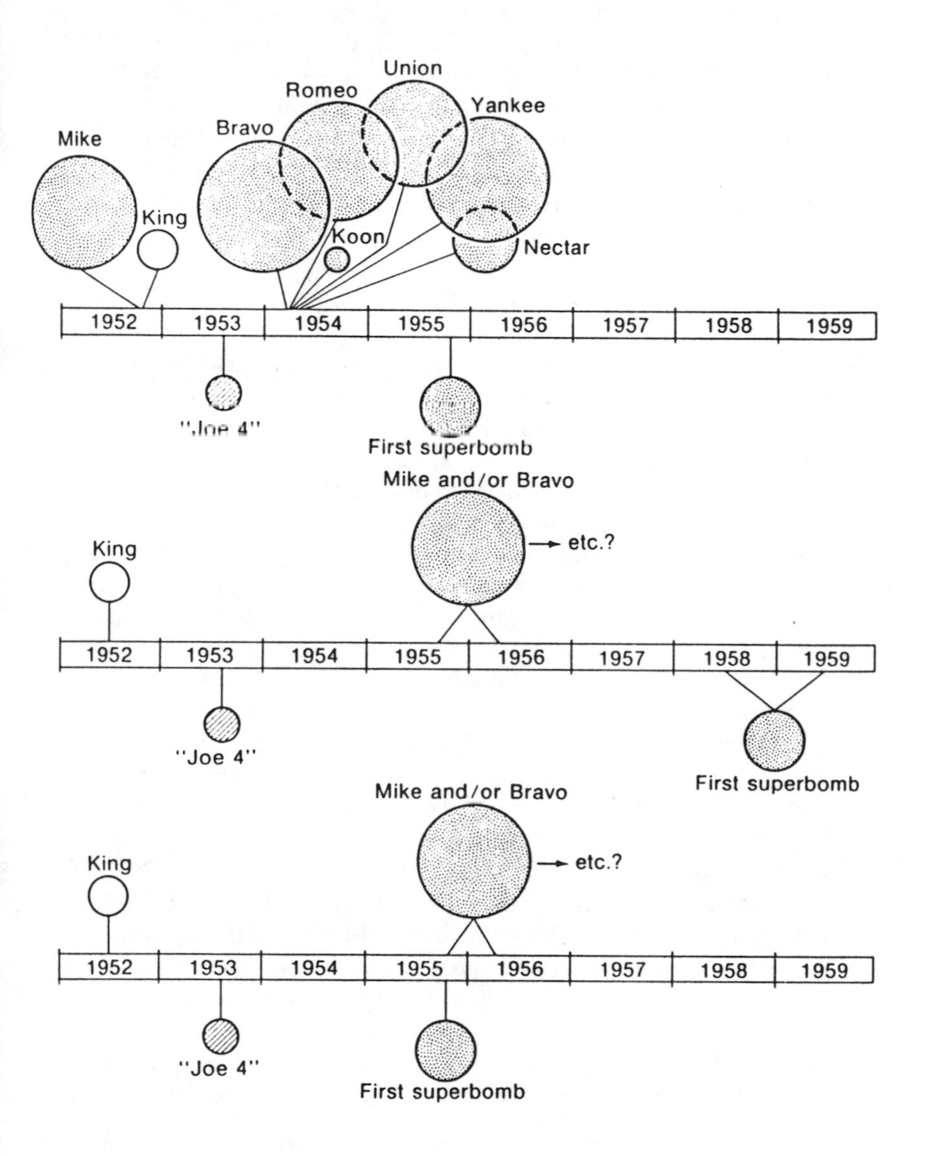
Union
Romeo
Yankee
Bravo
Mike
King
Koon
Nectar
1952
1953
1954
1955
1956
1957
1958
1959
First superbomb
Mike and/or Bravo
etc.?
King
1952
1953
1954
1955
1956
1957
1958
1959
"Joe 4"
First superbomb
Mike and/or Bravo
etc.?
King
1952
1953
1954
1955
1956
1957
1958
1959
"Joe 4"
First superbomb

a few hundred kilotons. But such a device would have had no real effect on the "balance of terror." In these alternative worlds the United States would surely already have tested the 500 kiloton "Super Oralloy" high efficiency all-fission bomb in November 1952, or more probably earlier, since the timing of Operation IVY was determined by the availability of the much more complicated Mike device. Therefore, the explosion of Joe 4 would have meant that the Soviet Union had only caught up with but not surpassed the United States insofar as the capability to produce enormous damage in a single explosion was concerned.

Then what would have happened?

From that point, the Russians might conceivably still have gone on to produce their multi-megaton explosion in November 1955, but I think it very probable that they would not have done so until very much later. In the actual world, they had the powerful stimulus of knowing from our November 1952 test that there was some much better, probably novel, way of configuring hydrogen bombs so as to produce very much larger explosions (twenty to thirty times as large) than the one that they demonstrated in their August 1953 experiment. Moreover, a careful analysis of the radioactive fallout from the Mike explosion may well have provided them with useful information concerning how to go about it. Writings and discussions by and about Igor Kurchatov make very clear his special interest in the fallout from U.S. nuclear tests and the information that could be learned from it, and Oppenheimer, in fact, had anticipated the usefulness of fallout for exactly this purpose. When asked in 1954 why he had in 1952 suggested there might be a benefit to the United States in an indefinite postponement of the Mike test, he said, "We[2] thought they would get a lot of information out of it[2]." In the hypothetical world where we would have followed the Oppenheimer–Lilienthal advice, that stimulus and that information would have been absent. And a comparison of the way Soviet and U.S. technology advanced during that period makes it seem very likely that there would have been a very much longer delay—probably

[2]"We" refers to a panel on disarmament reporting to the secretary of state, and including Vannevar Bush, Allen Dulles, John Dickey, and Joseph Johnson as its other members.

some years—before they took that big and novel a step without such stimuli and information. Therefore, in the most probable alternative world the first Soviet superbomb test would have been delayed until well after the first American superbomb test (as will be shown below, to 1957 or 1958), while in the worst plausible alternative world it could have occurred just when it did in the actual world, that is, in August 1955.

But what about the United States in the meantime?

We would have known immediately that the August 1953 explosion was partly thermonuclear, and that it was "many times" as big as their previous explosions. If we assume that following that Soviet test, the American program in our hypothetical world would have gone along just as it did in the actual world following Truman's 1950 decision, then we would have produced the Mike explosion in April 1956. However, a duplication of those earlier events at this later time under these different circumstances is very unlikely. Again, any analysis of U.S. reactions to Soviet technological advances shows that the detection of the August 1953 event would have resulted in the initiation of a really large, high-priority American program to produce a bigger and better thermonuclear device. Such a program would undoubtedly have been larger and would have had broader support than the one we actually did mount in the spring of 1950. Moreover, the general scientific and technological situation in which an H-bomb program would have been embedded in 1953 would have been significantly different from the actual one in 1950. For one thing, the kind of theoretical work in progress on the super before Truman's decision would have continued and would have provided a more solid base from which to launch a crash program. In addition, the booster program would presumably have continued along the path already set for it in 1948 (at which time a test of the principle in the 1951 operation was already being contemplated) and therefore in 1953 there would have been available some real experimental information concerning the thermonuclear reactions on a smaller scale. And last, but not least, there had been great progress in computer technology between 1950 and 1953. When the real MIKE was being planned, fast electronic computers such as the MANIAC and the first UNIVAC were either not quite operating, or were in the early part of their operating career. By a year or so later they were in full running order and much experience had been gained in their use, so that they would have been much more effective in connection with our hypothetical post-Joe 4 crash program. For all these reasons, it is most plausible to assume we would have arrived at something like the Teller–Ulam design for a multi-megaton superbomb either in the same length of time, or much more likely, a somewhat shorter period, say sometime between September 1955 and April 1956.

These dates bracket the actual date when the Soviets arrived at roughly the same point in the actual world. However, a few months' difference either way at this stage of the program would not have been meaningful. It takes quite a long time, several years typically, to go from the proof of a prototype to the deployment of a significantly large number of weapons based on it. Differences in production capacity and capability would have played a much more important role than any small advantage in the date of the first experiment, and such differences as then existed surely favored the United States. Hence, even if the Soviets had gone directly to their November 1955 design, which is the so-called worst plausible alternate world, (case 3 in Figure 1) it would not have resulted in upsetting the nuclear balance. Moreover, as already stated, I believe that in the most probable alternative world (case 2 in Figure 1) the date the Soviets would have arrived at that stage would have been delayed until well after the first very large U.S. Mike-like explosion showed them there was a better way. Hence, most likely, the United States in this hypothetical world would still have had a substantial lead.

Thus, the common notion that has persisted since late 1949 that some sort of disaster would have resulted from following the Oppenheimer–Lilienthal advice is, in retrospect, surely wrong. At worst, both countries would have learned how to ignite the large hydrogen bomb at roughly the same time. Moreover, even if by some unlikely quirk of fate they had arrived at that knowledge ahead of us, our very large stock of nominal fission bombs, plus the 500 kiloton all-fission bomb for those relatively few cases where it would have been appropriate, would have very adequately assured our national security. Much more likely, we would still have had the large superbomb well ahead of them in this alternative world just as we did in the actual world. And at best, although improbably, we might even have stopped the superbomb in time.

The GAC's Advice on the Super

The history and the conjectures about possible alternative pasts presented above show that Oppenheimer, Conant, Fermi, Rabi, and the others were right in their advice about the super, and they were right for the right reasons. They had correctly assessed the relative technological state of affairs, they had correctly judged the margin of safety inherent in it, and they had correctly projected our ability to catch up sufficiently rapidly if that should become necessary. American national security did not require the initiation of a high-priority program to develop the super. It was therefore entirely appropriate to attempt to use the first Soviet nuclear explosion as a lever for reopening the whole question of nuclear arms control instead of letting it serve simply as a stimulus—or an excuse—for the development of vastly more murderous bombs. Lilienthal and the majority of the commissioners were right in accepting the GAC's views: they had understood what the GAC was trying to say, and they had agreed with the way the GAC had weighed the possibilities.

At no point did the authors of the GAC report say or even imply that following their advice would necessarily prevent the development of the super by someone else. Rather, they said that the United States need not and ought not take the initiative, and they clearly and correctly recognized that our forgoing it was a necessary condition for persuading others to do so.

The authors of the report could not, of course, predict the details either of the chronology or of the alternative history propounded above, and they did not try to do so, but they could and did correctly assess the general situation and the limits of the probable futures inherent in it. The large rate of production of fissile material already in effect, the planned expansion in this rate, the resulting very large stock of fission weapons forecast for the early and mid-1950's, and the existence of an entirely adequate means for delivering these weapons guaranteed that even the sudden surprise introduction of a few superbombs by the Soviets could not really upset the balance. This situation was reinforced by the projection, proved correct in the King shot, that if need be, the power of the World War II fission bombs could be multiplied fifty or perhaps even a hundredfold (that is, up to the megaton range) simply by more astutely employing the techniques and materials already known and available.

Moreover as Fermi and Rabi implicitly projected in their minority addendum, the difficulty and novelty of the superbomb program would require preliminary large-scale experiments, and such experiments would provide adequate warning that a serious program was under way. Such experiments did, of course, actually occur in both programs. In the American program, the pioneering George shot preceded the experimental Mike explosion by eighteen months and the more practical Bravo by thirty-four months. In the Soviet program the relatively small Joe 4 explosion preceded the more nearly super-sized November 1955 explosion by twenty-eight months. In each case the radioactive debris could give evidence that thermonuclear reactions had taken place. Specifically, in the case of Joe 4, AEC Chairman Strauss was able to confirm promptly Premier Malenkov's claim that it included thermonuclear reactions.

Truman's Actions

The foregoing facts and arguments are, I think, sufficient to show that the GAC's advice was sound and right, but they may not be enough to show unequivocally that Truman should have taken it. The president, unlike the commissioners and their advisors, had to take into account a broader array of information and political ideas than those discussed in detail in this book. The overall intensity of the cold war was increasing, Mao and Stalin had proclaimed the Sino–Soviet bloc, and many important Republicans were withdrawing or modifying their support of the bipartisan foreign and military policies which had been in effect since World War II had begun. Moreover, as the fall of 1949 wore on, and the arguments about the super began to leak out from behind the curtain of secrecy, those opinions favoring the super were, in the overall cold war context of the time, both simpler and more widely persuasive than those opposing it. There can be little doubt that such congressional and public opinions as there were, were coming down strongly on the side of making a strong and virile response to Joe 1, and building the super seemed to be just the sort of thing to keep the Russians in their proper place. Truman, being a professional politician and therefore skilled in the "art of the possible," could therefore have concluded that rejecting the super and running even a small risk of being second best was simply too difficult an alternative. Moreover, the first and key decision in the chain that led to the development and production of the superbomb, that is the decision made on January 31, 1950 (see chapter 4), was based on the advice of the Special Committee of the NSC charged with studying the matter. That advice in turn called for a "minimal" decision (as Warner Schilling put it in his article aptly entitled "The H-Bomb Decision: How to Decide Without Actually Choosing"[3], namely, a decision to "continue" programs already going on to determine the feasibility of the super, and to study further its security and international implications. This decision did lead very quickly, and it seems inevitably, to the other subsequent decisions to produce and stockpile the super, and it was, as pointed out above, correctly taken to be a victory for those favoring an all-out superbomb program. Indeed, the decision to proceed with preparations for "quantity production"[4] followed the basic decision to "continue" the research program by only 6 weeks, a period much too short to allow the introduction of any pertinent new data. And very importantly, the recommendations leading to the first decision were fully supported by the committee members responsible for international relations (Secretary of State Acheson) and national defense (Secretary of Defense Johnson), and the only reservations were expressed by the one committee member *not* responsible for those elements of national security policy, namely Lilienthal, the chairman of the AEC. But, be all that as it may, it now seems clear to me in retrospect that Truman *should* have taken the advice; he should have held back on initiating the development of the super while making another serious try to achieve international control over all nuclear arms, including especially the super. The benefits that could have derived from forestalling the super altogether were incalculable; the odds of succeeding in doing so were small, but so were the risks in trying. It was certainly one of the few opportunities, and indeed as Lilienthal said then, it may have been the last good opportunity to base American foreign policy on something better than reliance on weapons of mass destruction, or as we would say it now, on the prospect of "mutual assured destruction." There can be no doubt that Truman would have found it very difficult to promulgate and conduct a policy of thermonuclear forbearance, but it might well not have been impossible. As the GAC urged, the most essential elements of the debate could have been made generally available, and the fact that several of the very few policy-oriented persons who were fully informed also came down on the side of restraint (for example, Lilienthal and Kennan) means that at the very least an attempt to sell such a position would not have been utterly hopeless.

Technical Predictions in the Report

In the course of presenting its admonition not to proceed with the crash development of the super, the GAC made certain very specific predictions about it. An examination of these predictions shows that they stood the test of time fairly well and throws further light on the accuracy of the GAC's assessment of the overall situation.

In their discussion of the superbomb, the GAC members said that "an imaginative and concerted attack on the problem has a better than even chance of producing the weapon within five years." Four years and four months later on March 1, 1954, Bravo, the first practical U.S. thermonuclear weapon, was tested at Bikini. Given the unknowns and uncertainties that existed at the time, that is truly a remarkably accurate prediction. They went on to say that "once the problem of initiation has been solved, there is no upper limit . . . except that imposed by requirements of delivery." That also seems to be the case. The largest bomb so far exploded (by the Russians in 1961) is said to have produced some 58 megatons—four times the size of Bravo—and there is every reason to believe bombs could indeed be made indefinitely larger than that.

The report also said that there "appears to be no experimental approach short of an actual test which will add to our conviction that a given model will or will not work" and that "many tests may be required before a workable model has been evolved." History has borne out the first part of this prediction. Twenty-five years later there were still no known ways to ignite thermonuclear fuels on a laboratory scale.[3] The second part of the prediction turned out to be less accurate. The number of U.S. tests needed to develop and check out a bomb were three: George, Mike, and Bravo. In the Soviet case it took only two, but they had one invaluable datum not available to us; namely, the sure knowledge that both small (George) and large (Mike) thermonuclear explosions were really possible. These numbers are very probably smaller than the "many" the GAC had in mind, but even so they were in each case sufficient to provide the other side with an early warning that thermonuclear work was in progress.

Another interesting and perceptive technological prediction is contained in the report's statement about ". . . very grave contamination problems which can easily be made more acute . . . by surrounding the deuterium with uranium." The very high levels of radioactive fallout associated with large hydrogen bombs do in fact derive from such use of uranium (that is, from ordinary uranium not specially enriched in the easily fissionable isotope U-235). The discussion of the fallout produced by very large hydrogen bombs presented in the AEC's book *The Effects of Nuclear Weapons* is based on the assumption that approximately one half of the energy of such explosions and virtually all of their

[3]To be precise, recently some thermonuclear reactions are reported to have taken place in laboratory experiments directed toward the eventual production of electric energy from micro-miniature thermonuclear explosions, but the level of these reactions and the external circumstances in which they took place were such that these results are not, for now at least, relevant in the context of this story. See, for instance, R. R. Johnson, "Observations of D-T Neutron Production from Laser-Driven Implosions," a paper given at the Albuquerque meeting of the American Physical Society Division of Plasma Physics, October 28–31, 1974.

fallout is derived from the fission of uranium. The very first test of a practical superbomb, the Bravo test at Bikini in March 1954, produced a blanket of fallout which evidently did contribute to the death of one innocent bystander (the radioman of the *Fortunate Dragon*, a Japanese fishing ship), and which came within a hairs breadth of killing some hundreds of Marshall Islanders living on two nearby atolls. This fallout accident in turn provided the initial spark behind the movement to ban nuclear weapons tests which ultimately led to the Partial Nuclear Test Ban Treaty of 1963.[4]

The GAC's Advice on the Booster Program

The GAC report strongly recommended going ahead with the booster program, subject only to the outcome of the then forthcoming 1951 Eniwetok tests, and it also recommended the development of reactors specifically designed to produce the tritium that its authors thought would be needed for that purpose. It is clear, therefore, that the GAC was not trying to stop thermonuclear work in general; it was not trying to forbid the exploration of the basic features of this novel but fundamental process; it was not, as it is sometimes charged, trying to outlaw some part of the fabric of science or to prevent the acquisition of new knowledge. Rather, the GAC was trying to inhibit only the development of the super, a device which from the very beginning had as its sole very specific purpose increasing the power of nuclear weapons one thousandfold. Their proposals were, therefore, consistent with what seems to be possible in Western civilization. It has not proved possible to control men's "thinking," or to prevent the acquisition of new fundamental knowledge.[5]

[4]For completeness it should be added that the GAC's predictions were in error with regard to the amount of tritium that would be needed to make a super work. However, the figures for both the amount the GAC predicted (note the asterisks in the Appendix) and also the amount actually used are still classified, and so no useful detailed comments can be made at this time. It can be said, however, that the main argument did not hinge on this estimate.

[5]It is of course sometimes possible for governments to control the rate at which new knowledge is acquired through the power of the purse, especially in those fields, such as high energy nuclear physics, where extraordinarily expensive pieces of apparatus are required. Even so, and generally speaking, such controls only determine the rate of scientific progress and hence they only affect when and not whether new knowledge will be acquired.

It has, however, on occasion been possible to stop or at least slow down the practical applications of new knowledge to specific devices or purposes (well-known examples are the supersonic transport, and both biological and chemical warfare) and that was all that the GAC was trying to do in this instance.

The GAC's Advice on Fission Weapons

In addition to its principal and most famous admonition not to proceed with the super, the October 1949 GAC report makes comments on several of the other then current proposals about how to respond to the first Soviet explosion. In the main, these proposals were pointed towards increasing the numbers and types of fission weapons. Specifically, the report in effect endorsed the commission's proposals for increasing the production of ore, expanding the facilities (isotope separation plants and production reactors) for converting it into nuclear explosives, and developing a greater variety of nuclear bombs for tactical purposes.

What we now (1975) know about the early Soviet nuclear weapons development and production programs makes it clear that even these actions were quite unnecessary, at least insofar as they were responses or reactions to the first Soviet explosion. Indeed, only a few years later it became obvious to everyone that the United States was accumulating excessive amounts of fissile materials. President Eisenhower, in his famous "Atoms for Peace" speech at the United Nations in 1953 made proposals that would eventually call for transferring large amounts of U-235 from military to peaceful purposes, and in 1956 he actually transferred 40,000 kilograms (enough to make several thousands of nuclear bombs). And even with such diversions to other purposes, the actual production rate of fissile materials was actually cut back in the late fifties and early sixties. However, the GAC's proposals to expand the production and widen the use of fission bombs were based on American estimates of the overall Soviet military capability, very largely conventional but now augmented by nuclear weapons, and on American perceptions of Soviet intentions, especially as these might be modified by the increased national self-confidence that the possession of a nuclear capability seemed to bring. It must be recalled that only the year before the Soviets had placed a military blockade around Berlin and had engineered a coup in Czechoslovakia under the direct protection of the Red Army, and that at the time the Communists were eliminating the last vestiges of opposition in the mainland of China and could logically be expected to support their Soviet allies in any new military adventures that might be in the wind. While none of this has any necessary connection with the development of the superbomb, it was plausible for AEC officials to propose and for the GAC to endorse the further development and deployment of tactical nuclear weapons as one means of coping with whatever other expansive moves might follow those of the last few years. In this connection, it makes no difference whether the revisionist historians are right or wrong about where the basic blame for the cold war lies, the perceptions of the AEC officials and the GAC members were based on the objective situation as described above, and that is what governed their advice.

Therefore, while with the full benefit of hindsight these expansions in production and in research and development on tactical and defensive versions of fission weapons can also be classified as overreactions, a simple, concrete rationale could then be given in favor of them, and they were not entirely out of proportion to their purported stimulus in the way that the proposed program to develop the super was.

Chapter SEVEN

Other Consequences

In addition to setting in motion processes which produced both the American and the Soviet superbombs, the first Soviet A-bomb test in 1949 and the American determination to respond to it also had other major consequences. Among the most important were (1) the conduct of a series of special *ad hoc* studies which both stimulated interest in the application of nuclear weapons to new purposes and in new situations and also at the same time served to raise doubts in some minds about the motives of Robert Oppenheimer, (2) the establishment in the United States of a second nuclear weapons laboratory, (3) a vast proliferation in the numbers and types of fission weapons in the U.S. arsenal, and (4) the 1954 security hearings in which Robert Oppenheimer was stripped of his security clearance and removed from the last of his positions of official influence.

The *Ad Hoc* Special Studies

The Soviet A-bomb and the ensuing activities in the United States brought about the initiation of a number of very important studies concerning the relationship between these events and various elements of American foreign policy, defense policy, and military strategy.

As reported earlier, when Truman issued his January 31, 1950, directive to go ahead with the hydrogen bomb program, he appended to it a letter to the secretaries of state and defense instructing them to make a complete review and reassessment of U.S. foreign and military policy in the light of the fall of China, the Soviet A-bomb, and the prospective U.S. H-bomb. This instruction fell on fertile ground. Admiral Souers, secretary of the NSC, had already recommended such a review only weeks before, and the NSC had concurred in that recommendation. Similarly, in the State Department, both George Kennan, then director of the Policy Planning Staff (PPS), and Paul Nitze, an assistant, had become convinced that existing plans had been made obsolete by the recent events, and had initiated discussions with Defense Department planners for the purpose of generating new plans and policies. As Kennan later put it, there was[1]

> unclarity in the councils of our Government as to the reasons why we were cultivating and holding these weapons. The unclarity revolved around this question. Were we holding them only as a means of deterring other people from using them against us and retaliating against any such use of these weapons against us, or were we building them into our military establishment in such a way that we would indicate that we were going to be dependent upon them in any future war, and would have to use them, regardless of whether they were used against us first.

Twenty-five years later, this "unclarity" still bedevils military planning in the U.S. (especially with respect to the use of nuclear weapons in a European War).

This concatenation of events promptly led to the establishment of the special study group that produced the report now known as NSC 68[2]. Secretary Acheson named Nitze, who had just succeeded Kennan as director of the Policy Planning Staff (PPS), to be chairman of the study. Several other members of the PPS also became members of the study committee. Defense Department members included Johnson's assistant for foreign affairs,

Major General James Burns, Ret. Other Defense Department representatives were Major General Truman Landon for the Office of the Joint Chiefs, Najeeb Halaby (Burns' deputy), and Robert LeBaron, chairman of the Military Liaison Committee (MLC). The special working group also received advice from a wide spectrum of outside experts, including Oppenheimer, Lawrence, and Teller.

Nitze started with the strongly held opinion that the U.S. response to recent events was inadequate, and the study, which took only six weeks to complete, confirmed his views. The report concluded that while the Soviets did not have a master plan for world conquest, they were determined to consolidate their power over their satellites, and they did have a desire for world hegemony, which was bound to lead to confrontations with other major powers. The report presented several alternatives, but it clearly favored across-the-board rearmament of the U.S. and its allies. The study committee did not estimate the cost of such a program of rearmament, but it seems clear that the State Department members were thinking in terms of $35 billion to $50 billion. The Defense Department members, on the other hand, supported Secretary Louis Johnson's budgetary views and seemed to be thinking in terms of very much less.

On April 12, 1950, President Truman referred the report to the NSC (which then gave it the file number NSC 68), and asked that the staff, augmented by experts from other agencies, work out the details of the forces needed to meet the goals set forth in it. Estimates of cost ranged up to $50 billion, but defense representatives maintained such figures were meaningless without further political considerations. It was also clear that neither the president nor his defense secretary would accept any such figure. These arguments were still going on when the Korean War broke out and presented its own rationale and justification for rearmament. Further modifications of NSC 68 did provide some guidance for the path rearmament followed (and helped to set the higher post-Korean arms levels) but it seems clear the rate and scale of the buildup were determined mainly by the war. In mid-1952, President Truman asked Dean Acheson, Robert Lovett, and Averill Harriman to review the matter again (in effect, to revise NSC 68), and to prepare a paper for the next administration. They in turn assigned the preparation of the paper to a group which again included Nitze. This new paper received the file number NSC 141. The Eisenhower administration ultimately rejected the report, finding it too expensive and too specific, and preferring naturally enough to make its own overall reviews of the matter.

Several of the persons involved in the preparation and subsequent analysis of NSC 68 had long subsequent careers related to the arms race. Paul Nitze later served as assistant secretary of defense for International Security Affairs, secretary of the navy, deputy secretary of defense in the Kennedy and Johnson administrations, and as a member of the U.S. delegation to the Strategic Arms Limitation Talks (SALT) negotiations during the Nixon administration. Richard Bissell, who was one of the experts loaned to the NSC staff to study the implications of NSC 68, and who also worked on NSC 141,[1] later became a deputy director of the Central Intelligence Agency (CIA). While with that agency, he was deeply involved in the U-2 program, the reconnaissance satellite program, and the planning for the Bay of Pigs. Still later he was president of the Institute for Defense Analyses (IDA), a nonprofit corporation involved mainly in making technical and operations analyses for the Joint Chiefs of Staff and other high echelons of the Defense Department.

In addition to the very broad NSC 68 study, the Soviet A-bomb and the ensuing events led to the establishment of the Long Range Objectives Panel in the Department of Defense and the initiation of a number of other more narrowly focused *ad hoc* study groups. These included the Air Defense Systems Engineering Committee (ADSEC); Projects Charles, Vista, and East River; the Lincoln Summer Study; and the Disarmament Advisory Panel of the State Department.

Of all these, only ADSEC came into being in 1949. It was created by the air force on the advice of the Air Force Science Advisory Board for the purpose of reexamining the nation's air defenses in the light of the Soviet bomb. The chairman of ADSEC

[1] He was then an economist on the staff of the Economic Cooperation Administration.

was George Valley, later one of the air force's chief scientists. Committee membership included Allen Donovan, C. Stark Draper, and Guyford Stever.[2] Most of the members of the committee were in the Cambridge area, and the committee conducted its work there, frequently in conjunction with the Air Force Cambridge Research Station. The committee quickly concluded that, obviously enough, the United States was fast becoming more vulnerable to enemy attack than it had ever been. After a year's work, they concluded that an effective air defense was feasible, and they outlined the technological means for accomplishing it. Ultimately, the report of ADSEC led first to Project Charles (another study) and then to the creation of M.I.T.'s Lincoln Laboratory.

The other special study projects cited above all owe their origins as much or more to the onset of the Korean War and the added burden of fears and concerns that war brought to American policy makers and strategic planners, as they do to the shock of the Soviet atomic bomb and the decision to go ahead with the superbomb. However, they are important in the immediate context because Robert Oppenheimer participated as a member or consultant in all of them, and he used his participation to further his own ideas about nuclear weapons and their proper role in foreign affairs and military strategy.

The Long Range Objectives Panel was established in late 1950 to examine once again the relationship of nuclear weapons to foreign policy, and military tactics and strategy. Oppenheimer was appointed chairman of the panel. Robert LeBaron, who is said to have proposed him for the post, knew very well Oppenheimer's general views on the superbomb, but reportedly believed that they would be effectively counterbalanced by the contrary views of others. Similarly, Oppenheimer himself invited Luis Alvarez, whose differing views were well known, to join the committee for that reason. Other members were Robert

[2]Donovan later was a key figure in the early development of the Ramo-Wooldridge Corp. and the establishment of the Aerospace Corp., and for almost thirty years has played a leading role in technical direction of the U.S. missile and space programs. Draper was for more than twenty years the leader of most U.S. inertial guidance development work. Stever was another air force chief scientist (1955–1956) and later (1972) director of the National Science Foundation.

Bacher, formerly a member of the AEC; Charles C. Lauritsen, a physicist, rocket expert, and longtime advisor to the highest levels in the navy; Mervin J. Kelly of the Bell Laboratories; W. G. Whitman of the Research and Development Board; Army General K. D. Nichols; Air Force General Roscoe C. Wilson; Admiral W. S. Parsons and General James McCormack, head of the AEC's Division of Military Applications.[3] The committee's report cited the Korean War as evidence that limited wars were possible and that general war with Russia could happen. It visualized possible uses for nuclear weapons in tactical situations and for defensive purposes, in both big and small wars. It emphasized the importance of weapons development programs whose goals were to reduce the dimensions of fission weapons and to increase the efficiency of use of fissile material. It said that the feasibility of thermonuclear weapons could not be established without further theoretical analysis, and it added,

> In fact, we believe that only a timely recognition of the long range character of the thermonuclear program will tend to make available for the basic studies of the fission weapon program the resources of the Los Alamos Laboratory[3].

Edward Teller complained to Alvarez shortly after the committee issued its report in January 1951 that this last conclusion was being used by the Los Alamos administration to slow down the thermonuclear program[4].

The reports of this and other special committees are still in the main secret, and so of course are the details of Oppenheimer's contributions to them. However, much of what went on was leaked in some officially inspired anti-Oppenheimer articles in 1953 (see the ZORC affair below) and came out in testimony at the Oppenheimer security hearing in 1954.

In general, Oppenheimer continued to stress the views that probably more and surely better atomic bombs were needed, that the superbomb was of doubtful military value, and that the pro-

[3]David Beckler served as secretary to the committee. He subsequently served for two decades as chief of staff of the President's Science Advisory Committee.

grams for the development of better A-bombs on the one hand and the development of the super on the other hand were seriously competing with each other for scarce resources—persons, facilities, and money—as well as for sufficient attention from the highest military and governmental authorities. Specifically, he seems to have advocated that everyone involved—including military planners in the Pentagon and the weapons designers at Los Alamos and Sandia—focus attention on the design and development of nuclear and conventional weapons for use in two particular situations: the defense of Western Europe against direct invasion by the Russians, and the defense of North America against air attack.

Among other controversial elements in his views was the notion that all three services, not just the air force, should be involved in planning, designing, and deploying such nuclear weapons. Moreover, as a sort of counterpoint to these positive opinions about tactical and air defense weapons, he continued to oppose on both ethical and practical grounds what was then the dominant view in the air force, namely that nuclear weapons should mainly be reserved for strategic bombardment purposes, and the more powerful the bomb the better for those purposes.

Oppenheimer's overall view of the situation during that period is available in a 1953 article he wrote for *Foreign Affairs*[5]. He later said[6] this article embodied three of the five recommendations made by the State Department's Disarmament Advisory Panel, of which he was a member (the other two involved specific diplomatic moves), and he then summarized those recommendations as follows: "that the people of this country be given a better understanding of the dangers of the atomic arms race, that we attempt either through administrative practice or through revised legislation to work more closely with our allies on problems having to do with the offensive and defensive aspects of large weapons, and three, that we take further measures for continental defense as a supplement to a striking capability." In that article he presented his well-known analogy of the "two scorpions in a bottle, each capable of killing the other, but only at the risk of his own life." In thinking about what to do about our situation, he said,

> There are three things we need to remember, three things that are very sharp. It is perilous to forget any one of them. One is the hostility and the power of the Soviet. Another is the touch of weakness—the need for unity, the need for some stability, the need for armed strength on the part of our friends in the Free World. And the third is the increasing peril of the atom.

He also spoke favorably of tactical types of nuclear weapons, and criticized the artificial limitations that might be placed on planning for their use by the uninformed opinions of our allies. Perhaps most importantly, he called for more candor on the part of the authorities and for the widest possible discussion of the issues:

> [We must have] candor on the part of the officials of the United States Government to the officials, the representatives, the people of their country. We do not operate well when the important facts, the essential conditions, which limit and determine our choices are unknown. We do not operate well when they are known, in secrecy and in fear, only to a few men.
>
> The political vitality of our country largely derives from two sources. One is the interplay, the conflict of opinion and debate, in many diverse and complex agencies, legislative and executive, which contribute to the making of policy. The other is a public opinion which is based on confidence that it knows the truth.
>
> Today public opinion cannot exist in this field. No responsible person will hazard an opinion in a field where he believes that there is somebody else who knows the truth, and where he believes that he does not know it.

Oppenheimer's role in all these various panels and committees came to be seen by some in sinister terms. When asked why he let Oppenheimer write a report (for the Long Range Objectives Panel) that was purportedly being used to undercut the superbomb program, Alvarez explained that he simply had been outsmarted by Oppenheimer and had not until just then realized what the implications of the report were. "Oppenheimer was, in effect, practicing Political Science 4B, on the post-graduate level, while I was still on Political Science 1A, the freshman course[7]." Air

Force General Roscoe Wilson was more disturbed. He later testified[8]:

> The panel contained some conservative statements on the possibility or the feasibility of an early production of a thermonuclear weapon. These reservations were made on technical grounds. They were simply not challengeable by the military. They did, however, cause some concern in the military.
>
> It is hard for me to explain this, except to say that most of us have an almost extravagant admiration for Dr. Oppenheimer and Dr. Bacher as physicists, and we simply would not challenge any technical judgment that they might make. But I must confess, and I find this exceedingly embarrassing, sir, that as a result of this panel and other actions that had taken place in the Committee on Atomic Energy that I felt compelled to go to the Director of Intelligence to express my concern over what I felt was a pattern of action that was simply not helpful to national defense.

The committee referred to in the foregoing quotation was one of the subcommittees of the Research and Development Board, another of the many interlocking groups Oppenheimer was either a member of or a consultant to during that period. General Wilson was especially upset over Oppenheimer's negative attitude toward a project whose goal was the development of a nuclear-powered airplane. Despite Oppenheimer's (and the GAC's) objections, the project went ahead anyway only to be cancelled in 1960 after the expenditure of a billion dollars. As it turned out, the project was grossly technologically premature. This was another instance where Oppenheimer and his colleagues had been right for the right reasons.

David Griggs, then air force chief scientist, was even more disturbed. He noted that Oppenheimer not only was cool to the Strategic Air Force, the superbomb and a second weapons laboratory, but that in addition, during Project Vista, he had proposed a three-way split in the allocation of fissile material at a time when the air force felt there simply wasn't enough for just its own needs. Griggs accused Oppenheimer of being a member of a cabal code-named ZORC (Z for Zacharias, O for Oppenheimer,

R for Rabi, and C for C. C. Lauritsen)[4] and engaged in nefarious anti-air force schemes.

In his testimony at Oppenheimer's security hearing in 1954, Griggs expressed his, and the air force's concerns in these words[9]:

> It became apparent to us—by that I mean to Mr. Finletter, Mr. Burden, and Mr. Norton, that there was a pattern of activities all of which involved Dr. Oppenheimer. Of these one was the Vista project—I mean was his activity in the Vista project, and the things I have already talked about. We were told that in the late fall, I believe of 1951, Oppenheimer and two other colleagues formed an informal committee of three to work for world peace or some such purpose, as they saw it. We were also told that in this effort they considered that many things were more important than the development of the thermonuclear weapon, specifically the air defense of the continental United States, which was the subject of the Lincoln Summer Study. No one could agree more than I that air defense is a vital problem and was at that time and worthy of all the scientific ingenuity and effort that could be put on it. We were however, disturbed at the way in which this project was started. . . .
>
> It was further told me by people who were approached to join the summer study that in order to achieve world peace—this is a loose account, but I think it preserves the sense—it was necessary not only to strengthen the Air Defense of the continental United States, but also to give up something, and the thing that was recommended that we give up was the Strategic Air Command, or more properly I should say the strategic part of our total air power, which includes more than the Strategic Air Command. The emphasis was toward the Strategic Air Command.

In summary, the proponents of more and better tactical and air defense nuclear weapons on the one hand, and the proponents of strategic weapons and the superbomb on the other, saw each other both as ideological threats and as serious competitors for

[4]This ZORC affair was first brought out in an article by Charles V. Murphy in *Fortune* magazine, May 1953, pp. 109–110. Murphy was an air force reservist who served occasionally in the Pentagon itself, and had excellent contacts and sources in air force headquarters. Z, O, R, and CCL all denied there was any such cabal.

what most of them considered to be an inadequate total supply of resources. However, as things actually eventually worked out, both sides got what they wanted, and the various different weapons developments actually complemented and stimulated each other. One of the factors that led to this unexpected result was the establishment of the second nuclear weapons laboratory at Livermore.

The Lawrence Livermore Laboratory

Three originally separate strands of events, each a part of the fabric of the superbomb controversy and the events flowing from it, eventually coalesced and led to the creation of a second American nuclear weapons laboratory at Livermore, California, in the summer of 1952. One of these strands was the determination of Ernest Lawrence and Luis Alvarez to involve themselves and their colleagues at Berkeley in some direct, useful, and important way in the American response to the first Soviet A-bomb. Another was the protracted conflict between Edward Teller on the one hand, and Norris Bradbury and his senior staff on the other, about how the Los Alamos laboratory might best go ahead on the H-bomb program. This conflict finally led Edward Teller to conclude that a second laboratory had to be established to do the job adequately. The third factor, was the happenstance that a small group of Berkeley scientists participated in the George experiment at Operation Greenhouse, thereby creating at Berkeley a small cadre of young men familiar with the details of thermonuclear weapons design.

On the very day the news of the first Soviet A-bomb became known to the public, Lawrence, Alvarez, Latimer, and others at Berkeley began to ponder the appropriate American response to that event, and to search for ways they themselves might participate in such a response. They discussed the matter among themselves, and they then traveled to other centers of nuclear research to learn the views of other scientists. Among the places they visited was Los Alamos, where they were particularly interested in learning more about Teller's ideas about the subject.

On the basis of these early explorations, they concluded that they should support Teller's proposals for an urgent, high-priority program at Los Alamos to develop a superbomb based on the fusion process, and that the Berkeley group should undertake the design and construction of a reactor which could produce a large excess of neutrons. Teller had explained to them that substantial amounts of tritium—a heavy radioactive form of hydrogen which does not occur naturally—might be needed in the development and manufacture of fusion (hydrogen) bombs, and they knew that the best way to produce tritium was in a reactor specially designed to produce a large excess of neutrons.[5] The GAC, in its famous meeting of October 1949, agreed that the design of such a reactor should be undertaken,[6] but it suggested that the program be carried out by the Argonne National Laboratory, which had very much more relevant experience.

Lawrence and Alvarez were at first disappointed at this turn of events, but they soon responded with an entirely new concept based partly on an idea of Winn Salisbury. It soon became known by its cover name: the Materials Testing Accelerator,[7] or the MTA. The basic idea involved a two step process: first, produce large quantities of free neutrons by brute force; and second, absorb these neutrons in suitable materials to produce any of several desired end products—tritium, plutonium, U-233 (another fissionable material suitable for bombs), or radiological warfare agents.

To accomplish the first step in this process, they proposed building an enormous particle accelerator capable of producing as much as an ampere of deuterons having energies of several hundreds of millions of volts. Such a device would consume some hundreds of megawatts of energy—about what a large reactor produces—and it would, they speculated, produce a somewhat

[5]All fission reactors produce huge numbers of neutrons, but a large proportion of them are needed simply to keep the fission reaction going, and others either escape into the reactor shield or produce plutonium by reacting with the U-238 which is normally present. The production of large amounts of tritium by the $Li^6(n,T)He^4$ reaction requires that there be excess neutrons beyond those involved in the foregoing processes, and that in turn required that a new style of reactor be designed specifically for that purpose.

[6]See Appendix.

[7]By analogy to the Materials Testing Reactor, which really did have that purpose.

larger number of available free neutrons than that same large reactor. Lawrence first asked Robert Serber and me to make theoretical estimates of the production of neutrons in such a device, and then asked me to check them experimentally. I had just received my Ph.D. in physics at Berkeley, and had stayed on at the laboratory as what today would be called a "postdoctoral fellow." My data revealed that suitably large numbers of neutrons[8] would be produced almost no matter what materials were used to construct the primary target for the deuteron beam.

To accomplish the second step in their novel process, they proposed to surround the primary target with a large secondary target lattice in which the free neutrons produced in the first step would be absorbed in a suitable fertile material. For example, if tritium were the desired end product, then the secondary target would be constructed so that most of the neutrons were absorbed by lithium-6 atoms. Or, if plutonium were desired, then the secondary target would be constructed so that most of the neutrons were absorbed by U-238 atoms. In the latter case, the secondary target would be much like the core of a nuclear reactor, but it would be in principle much easier to design and build. A reactor core must be designed with criticality in mind, that is, it must be very carefully designed to that the so-called neutron chain reaction will continue at the desired rate. The neutrons for the MTA secondary target, on the other hand, are supplied from outside, and so no such careful design is necessary. Much more important, the uranium in this secondary target can be uranium that had been depleted in U-235—that is, the uranium left over after the isotope separation plants and the plutonium production reactors had largely removed or consumed the U-235 it originally contained. This in turn meant that in principle the MTA made it possible to exploit the basic raw material uranium ore much more completely and efficiently than would otherwise be the case.

Partly because of these potential benefits and partly because of Lawrence's enthusiasm and his reputation as a scientist who "knew how to get things done," the development and construction of a prototype machine was authorized by Washington. There

---

[8] 5 to 10 neutrons per deuteron at 350 million volts.

was, as would be expected, some dispute about such a novel and expensive device, but the Korean War broke out while the project was still in its beginning phase, and that completely settled the matter for the time being. The Korean War was widely interpreted as a switch by the Communists from a program of world conquest by political subversion to a program of direct—if for the nonce distant—military action. Since a very large part of America's uranium supply came from overseas—mostly the Belgian Congo and South Africa—the worsening world situation was widely perceived as justifying such a draconian approach to making more complete use of our native supply of ore.

As was customary for such complex nuclear projects, it was decided to proceed by stages. As a first step, a prototype machine, suitable for checking out the basic design principles, would be built at a site convenient to the Berkeley laboratory. Later, but as soon as possible, a full-scale machine would be built at a site whose selection would be based on other factors, including the availability of sufficient land, operating personnel, and electric power, and invulnerability to surprise attack.

A search for a convenient site suitable for the prototype resulted in the selection of an abandoned World War II naval air station some forty miles southeast of Berkeley, at Livermore, California. The California Research Corporation, a subsidiary of Standard Oil of California, was awarded a contract to build the prototype machine under the scientific guidance of the UCRL group, which had by then expanded to include Wolfgang K. H. Panofsky, William Brobeck, and many other Berkeley physicists and engineers. It was recognized by all concerned that an industrial contractor would be necessary both to build and to operate a full-scale plant—already being planned for construction near St. Louis, Missouri—and the California Research Corporation[9] was being brought in at the prototype stage in order to gain the necessary experience as quickly as possible.

In September 1950, three months after the outbreak of the Korean War, a full-scale version of the MTA was tentatively authorized for construction at Weldon Spring, Missouri. Known

[9]Among those on the CRC staff was Kenneth Davis, who later became director of the AEC's Reactor Division. Fred Powell was president of CRC.

as the A-12, it was to produce one half of an ampere of 350 million volt deuterons. As then conceived, the accelerator itself was to be approximately 60 feet in diameter and 1450 feet long. Prior to that time, the largest similar machines, known in the physics trade as linear accelerators, were typically a few feet in diameter and some tens of feet long.

The Livermore prototype was actually a full-scale model of the front end of the A-12 machine, and an accelerator 60 feet in diameter and 87 feet long was actually built there. It took longer to build than anticipated, and it never did run well. The MTA turned out to involve too many too large technological steps to be taken successfully all at one time. In addition, at least in the view of the Berkeley group, the California Research Corporation personnel were too often too slow in understanding what the "real" problems were. Finally, and before all the "bugs" were worked out of the design, it became clear that the main original reason for building it—a threatened shortage of uranium ore—was no longer valid. Within just a few years after the AEC first anticipated the possibility of a serious shortage, it was discovered that simply raising the price of uranium inspired prospectors to discover very large supplies of ore in Colorado and Canada. As a consequence, the political basis of the MTA was also eroded away. As a result, the authorization for the full-scale Weldon Spring machine was canceled on August 7, 1952, and the Livermore model was shut down and its dismantling started in December 1953. The building the MTA was housed in, and some of the equipment used to support its operation, later were used in support of the "Sherwood program," whose purpose was the controlled production of thermonuclear energy[10].

In the meantime, in the spring of 1950, while I was still refining measurements of the potential neutron yield of the MTA, Luis Alvarez approached Hugh Bradner and me to tell us something about the expanding work on the superbomb at Los Alamos. He had, he said, recently been talking with Edward Teller about the matter, and it appeared that the project could use some assistance from scientific groups at other laboratories. Bradner and I promptly flew to Los Alamos to meet with Teller and others, and we quickly agreed to set up a special group at Berkeley to perform some diagnostic experiments on the first thermonuclear test explosion, then planned for the following spring as the George shot of Operation Greenhouse. The group, code-named the "Measurements project," consisted of about forty persons, roughtly half of whom were young physicists of the type we would now call "postdocs." In essence, we were to make experimental observations of certain physical phenomena as these unfolded during the first fraction of a microsecond of the thermonuclear explosion.

I recall that several different considerations strongly motivated and inspired me to participate in the hydrogen bomb program. One was my own perception of the growing seriousness of the cold war, much influenced by my very close personal student–teacher relationship with Lawrence. The Sino–Soviet bloc had just been formed; Stalin and Mao both said that it was monolithic and that its goal was world revolution. Another inspiration was the scientific and technological challenge of the experiment itself; it was to be the very first occasion in which a thermonuclear reaction took place on the surface of the earth, and we were to make complex observations extending over a period of less than a millionth of a second. Five years before, I had played a peripheral role in the Manhattan project. I had not participated in the Trinity test (of the first A-bomb) and I had only heard about it a week or so after it occurred. This time I was being invited to participate directly in the heart of the matter. Yet another strongly favorable consideration was my discovery that Teller, Bethe, Fermi, Von Neumann, Wheeler, Gamow, and others like them were at Los Alamos and involved in this project. They were among the greatest men of contemporary science, they were the legendary yet living heroes of young physicists like myself, and I was greatly attracted by the opportunity of working with them and coming to know them personally. Moreover, I was not cleared to see GAC documents or deliberations, and so I knew nothing about the arguments opposing the superbomb, except for what I learned secondhand from Teller and Lawrence who, of course, regarded those arguments as wrong and foolish. (I saw the GAC report for the first time in 1974, a quarter of a century later!)

I mention my own motivations because I think I understand them better than the motivations of others. I do so here only to explain, not to justify or to rationalize. I do so because I think I was not untypical, and because my reactions can be expected to have occurred in others in similar situations. Oppenheimer's[11] later reaction to Teller's explanation of the ideas about how to make an H-bomb—"technically so sweet that you could not argue about it"—is a related response in someone vastly more sophisticated than I was. Other reminiscences of similar situations also clearly express the excitement scientists and other humans commonly find in such huge history-making events[12].

Most of the preparatory work of our group was done in Berkeley. However, the pilot setup of our electronics gear required more room than was readily available at Berkeley, so for that we used some space in the former naval infirmary at the Livermore station. The California Research Corporation was already at work on the MTA project at the site, and its working relationship with UCRL made it natural and simple for it to provide us with all the necessary housekeeping functions.

During March and April 1951, most of the members of our special group moved out to Eniwetok Atoll, in the Marshall Islands, and we set up our equipment in its final form there, in the shadow of the George device.

On May 8, 1951, at Eniwetok Atoll the first thermonuclear test explosion on earth was successfully conducted. The tritium deuterium mixture[13] burned well, and the various diagnostic experiments—including that of our Berkeley group—were also successful in recording the various phenomena that accompanied the explosion.

Some of the members of our Berkeley group, after completing the analysis of their data, participated in the general post-experiment discussions and in some of the future planning sessions. No specific plans for further participation resulted from these discussions, however, and so the Berkeley group was disbanded and its members turned to various other projects, mostly pure research in high energy physics.

■ ■ ■

Edward Teller stayed on at Los Alamos for another six months after the George shot. The next major experiment—the Mike shot of Operation IVY—was to be based on the Teller–Ulam idea, and he participated directly in the determination of its basic configuration. In November 1951, he left Los Alamos and returned to the University of Chicago. He did so in part because he felt the remaining theoretical work that was still needed to get Mike ready could be done just as well without him—Bethe was already scheduled to be at Los Alamos during the final design period—but mainly because the ancient arguments between him and Bradbury over how to run the laboratory and the hydrogen bomb program continued to worsen. Thus, in the summer of 1951, only some months after he came up with the final, capping suggestion in the series of ideas that led to the invention of the superbomb, he concluded that the establishment of a second, independent laboratory was needed to exploit this new approach in a timely fashion. As Teller later put it[14],

> It was an open secret, among scientists and government officials, that I did not agree with Norris Bradbury's administration of the thermonuclear program at Los Alamos. Bradbury and I remained friends, but we differed sharply on the most effective ways to produce a hydrogen bomb at the earliest possible date. We even disagreed on the earliest possible date itself, on the timing of our first hydrogen bomb test. The dissension with Bradbury crystallized in my mind the urgent need for more than one nuclear weapons laboratory.
>
> I knew that science thrives on friendly competition, on the fostering of different points of view, and on the exchange of ideas developed in different surroundings. I knew, too, that a single group of scientists working together can easily become fascinated by special aspects of a development—to the neglect of other hopeful approaches. My conviction grew that the safety of our country could not be entrusted to a single nuclear weapons laboratory, even though that laboratory were as excellent as Los Alamos. This conviction was hardened by a growing awareness, as our work progressed at Los Alamos and our thermonuclear knowledge increased, that we were pioneering a big new field of weapons development. I began to doubt that one laboratory would be physically capable of handling all the work that had to be done. Weighing all of these ideas and circumstances, I came to the inescapable conclusion that at least two weapons laboratories, working in cooperation but also

> in the traditional American spirit of competition toward the mutual goal of adequate national defense, were vital to the future of the United States.
>
> I also concluded that I could advocate establishment of a second weapons laboratory most effectively if I were not associated with the existing Los Alamos Laboratory. So, regretfully, I left Los Alamos in November 1951, and returned to the University of Chicago.

Teller soon succeeded in persuading Gordon Dean, who had by then replaced Lilienthal as chairman of the AEC, to consider the matter. At Dean's request the GAC reviewed the idea. Except for Willard J. Libby, like Teller a professor at the University of Chicago, and then a new member of the committee, it opposed the idea on the ground that the establishment of a second laboratory would divert talent and resources from Los Alamos and thus slow down the overall program. Very probably, the opposition of the GAC to a second laboratory also importantly involved personal elements. Teller's claim that competition was a good thing was often expressed in terms which made it clear he felt the Los Alamos leadership was unimaginative, negative, and otherwise inadequate. It was equally clear that Teller felt much the same way about many of the members of the GAC itself, and so it is not surprising that the GAC supported Los Alamos and Bradbury against what they regarded as an unwarranted personal attack.

Teller also sought support of his ideas in the air force. The air force would be the principal user of the hydrogen bomb, and a number of persons at the top of the air force very quickly evinced great personal and institutional interest in the issues being raised. David Griggs, one of the founders of the Rand Corporation and just then chief scientist of the air force; James (Jimmy) H. Doolittle, a much respected retired general and a high level general consultant to the air force leadership; and General Elwood ("Pete") Quesada, the commander of the joint task force that had conducted Operation Greenhouse, all became strong partisans of Teller, and they helped him make further contact with higher air force officials. As a result, Teller and his ideas were warmly received and strongly endorsed by Thomas Finletter, secretary of the air force, and his special assistant for research and development, William A. M. Burden.[10] They, in turn, began to make moves toward establishing a second nuclear weapons laboratory under air force sponsorship, and in 1951 they actually did arrange to sponsor briefly some nuclear calculations Teller was doing at Chicago. For this particular work, they used the facilities of Project Chore, a minor army project that had been going on at the University of Chicago for some years and which could handle classified work. However serious their intentions concerning a full-scale second laboratory may have been, their actions greatly increased the pressure on the AEC either to do something itself or to see its monopoly in the field eliminated. During this period, Teller also was given the opportunity to brief Secretary of State Acheson, Secretary of Defense Robert Lovett, and Deputy Secretary of Defense William C. Foster. The very fact of these briefings, independent of their specific result, put further pressure on the commissioners.

The congressional JCAE again played a crucial role. The chairman, Senator McMahon, was mortally ill at the time, and so the staff director, William Borden, personally conducted on his behalf a campaign for a second laboratory—and against the GAC leadership and its views.

In the meantime, in late 1951, Thomas Murray, the AEC Commissioner most sympathetic to the idea of a second laboratory, got in touch with Ernest Lawrence to discuss the matter with him. Lawrence was already well known to be very sympathetic to the idea of expanding the thermonuclear program. He responded positively, and volunteered to study the matter further himself. Since I had been more deeply involved in the recent thermonuclear program than anyone else at Berkeley, Lawrence in January 1952 asked me what my views were. As a direct result of Lawrence's inquiry, I made a series of extended trips to Los Alamos, Chicago, and Washington, where I discussed the matter

[10]Not to be confused with William L. Borden, executive director of the JCAE staff. Burden, by profession a financier, has had a long involvement in government affairs, especially in national security matters. He was later ambassador to Belgium in the late fifties, and in the sixties and seventies he was chairman of the Board of Trustees of IDA.

with most of the people named above plus a few others, including Army General Kenneth Fields, then the AEC's director of the Division of Military Applications, and his deputy, Navy Captain John T. Hayward. I found the whole affair heady and exciting (I had just turned thirty), and I was readily persuaded to Teller's point of view. I reported to Lawrence that I, too, felt it would probably be useful to establish a second laboratory. The idea of doing so at the Livermore site was, for us, a natural one, and we suggested it immediately to AEC authorities.

That specific addition to the general idea changed the nature of the argument. A proposal to establish a second laboratory in existing facilities at Livermore as a branch of the UCRL, as compared to a proposal to simply establish one "somewhere" under an unspecified aegis, clearly meant much less initial expense and an immediate, if small, cadre of people ready to go to work right away. As a result, and as GAC Chairman Oppenheimer later recalled, the GAC and the AEC ". . . approved the second laboratory as now conceived because there is an existing installation, and it could be done gradually and without harm to Los Alamos[15]." As I recall it, Lawrence and Teller felt at the time that Oppenheimer himself was still really opposed to a second laboratory but that under the new circumstances he had no other choice. Even so, during that year I met with Oppenheimer at Princeton, and discussed the plans for the Livermore laboratory. He received me in a personally friendly fashion, but I cannot recall his being of any particular help.

The precise nature of the plans for the new laboratory, however, primarily reflected Lawrence's ideas about how to go about such things, and deviated considerably from Teller's views of what should be done. In essence, Lawrence firmly believed that if a group of bright young men are simply sent off in the right direction with a reasonable level of support, they will end up in the right place. He did not believe that the goals needed to be spelled out in great detail, nor that it was necessary that the leadership consist of persons who were already well known. Teller on the other hand, had become deeply suspicious of the intentions of the AEC leadership, and he therefore wanted something more nearly analogous to the 1943 plans for Los Alamos, that is, a plan for a laboratory that would be led by a large cadre of famous scientists and that would have a well-defined goal.

To complicate matters, during that spring Lawrence, then suffering from a chronic illness, spent much time away from Berkeley on long rest trips. As a result I was left pretty much on my own to draw up the specific plans for a second laboratory with nothing except the most general guidance from my immediate superior. However, as a result of ten years of close association, I both clearly understood and firmly agreed with Lawrence's approach to "big science," and I generated plans which he always warmly endorsed when he had a chance to review them.

Finally, and in close accordance with Lawrence's (and my) views of the matter, the AEC in June 1952 approved the establishment of a branch of the Berkeley laboratory at Livermore which would assist in the thermonuclear weapons program by conducting diagnostic experiments during weapons tests and other related research, but the question of how soon (or even whether) the Livermore laboratory would actually engage directly in weapons development was left open. The AEC's official planning document[16] described the mission of the Livermore laboratory this way:

> a. Development and experimentation in methods and equipment for securing diagnostic information on behavior of thermonuclear devices and the conduct of such instrumentation programs in support of tests of thermonuclear devices in close collaboration with the Los Alamos Scientific Laboratory.
>
> b. While the work authorized above is the immediate objective of this proposal, the Commission hopes that the group at UCRL (Livermore) *will eventually suggest broader programs of thermonuclear research* to be carried out by UCRL or elsewhere. (Emphasis added.)

Lawrence felt that that kind of statement of intentions provided an adequate base upon which to build a second weapons laboratory. I would have preferred something more concrete, but I was prepared to accept it as a place to start from. Teller, on the other hand, found the vagueness of the AEC's plans for the Livermore laboratory entirely unsatisfactory. As a result, in early July he told Ernest Lawrence, Gordon Dean, myself, and others that he would have nothing further to do with the plans for establishing a laboratory at Livermore. The Berkeley administration was prepared to go ahead anyway. However, at the insistence of Captain Hayward more than anyone else, intense negotiations were resumed among all concerned. Within days, these led to a firm commitment on the part of Gordon Dean that thermonuclear weapons development would be included in the Livermore program from the outset, and a renewed commitment on the part of Teller to join the laboratory.

The laboratory was launched in September 1952. I became the director, and the Scientific Steering Committee included Teller, Harold Brown, John S. Foster, Jr., Arthur T. Biehl, and a few others. Teller, because of his obvious special status, was given veto authority over the decisions of the committee, but otherwise had no formal authority. Brown was put in charge of the development of thermonuclear weapons at Livermore, and first Biehl and then Foster was put in charge of the development of improved fission weapons. There were some early problems in the administration of the Theoretical Division, but these were resolved on a temporary basis by making Richard Latter, a Rand physicist then temporarily on loan four days a week to the Livermore laboratory, the acting head of the division. After about a year, he was replaced by Mark M. Mills, who remained in that position until his death in a helicopter crash at Eniwetok in early 1958. These organizational arrangements, although they contained some peculiar elements, worked out very well, and none of the strained relationships that had surrounded Teller at Los Alamos developed at Livermore.

The relationship between the Livermore and the Los Alamos laboratory however was strained from the beginning, and rapidly grew worse. One of the causes of this state of affairs involved the question of who deserved the credit for Mike, the first large thermonuclear device. In November 1952, two months after the Livermore weapons laboratory was established, Mike was exploded during Operation IVY at Eniwetok. Based on ideas by Teller, Ulam, and many others, it was designed, built and tested by the Los Alamos laboratory. Los Alamos was aided in those tasks by a number of other institutions including the Naval Research Laboratory and the Matterhorn-B group at Princeton, but not by the Livermore laboratory, which then was completely absorbed in simply coming into being. Nevertheless, elements of the press commonly credited Teller and the Livermore laboratory with creating Mike. The reasons for this error were Teller's presence at Livermore, and an absurdly strict secrecy policy on the part of the AEC. The AEC, purportedly for security reasons, refused either to make or to allow any comment whatsoever on the origin and nature of Mike, not excepting even a simple denial of its being created by Livermore. This situation continued for almost two years, until after Shepley and Blair[17] brought out a somewhat distorted account of the matter, and Bradbury was finally allowed to respond to that by holding a press conference giving his version of history.[11] In addition, Robert Oppenheimer's clearance was lifted in late 1953, and hearings in the matter were held in the spring of 1954. Despite elaborate efforts to assert the contrary, the real stimulus for the removal of Oppenheimer's clearance was his opposition to the program for superbomb development. Teller, along with Lawrence, Alvarez, Latimer, and some others in the University of California, provided some of the most important arguments in support of the government's case. On the other hand, both the veterans of the wartime Los Alamos laboratory, as well as its current leadership, strongly supported Oppenheimer. As a result, relations between Livermore and Los Alamos, which were bad to begin with, became very much worse. All during those very trying times, the Los Alamos administration treated the Livermore leadership formally correctly, and provided some much needed technological assistance to the new laboratory, but some of the leading scientists at Los Alamos made it clear that Teller was personally unwelcome there. Since Teller was such a central figure at Livermore, this feeling inevitably affected the relationship between the laboratories.

On top of all that, the first Livermore tests—in Nevada in 1953 and at Bikini in 1954—went poorly. Considering the attacks on the qualitv of the Los Alamos leadership that were part of the

[11]Teller's summary of the history is given in *The Legacy of Hiroshima*, and in *Science*, February 25, 1955, p. 268.

arguments supporting the establishment of a second laboratory, it is not surprising that some Los Alamos scientists filled the air with horse laughs on those occasions. Despite all that, the Lawrence Livermore Laboratory[12] flourished and eventually produced its share of new weapons ideas and designs. From a budget of $3.5 million for its first year's operations (fiscal year 1953) and a staff of 698 at the end of its first year, the laboratory grew steadily. Five years later, in fiscal year 1958, during my last year as director, the staff passed the 3000 mark, and the budget reached $55 million. In another five years (fiscal year 1963) the staff passed 5000 and the budget reached $127 million. Shortly after that, the laboratory stabilized at a population of about 5500.

When I went to Washington in the wake of Sputnik in late 1957, I was succeeded as director of the laboratory by Edward Teller. He was in turn followed by Harold Brown (1960), John S. Foster, Jr. (1961), Michael May (1965), and Roger Batzel (1973), all of whom were members of the laboratory itself from its earliest years. Alongside those changes, Norris Bradbury remained the director of Los Alamos laboratory until his retirement in 1970. He was then succeeded by Harold Agnew, one of the younger wartime Los Alamos pioneers.

The Proliferation of Nuclear Weapons

In the fifties and early sixties the numbers, types, deployments, and end uses of the nuclear weapons in the American arsenal proliferated far beyond anything conceived of in the early nuclear years. The underlying causes of this remarkable proliferation were all intimately connected to the superbomb controversy, and the events that flowed from it. In no particular order they were (1) the establishment of the second weapons laboratory, (2) the special attention focused on tactical nuclear weapons by Oppenheimer and his colleagues, at least partly as an element of the struggle to contain the superbomb, (3) the increases in the rate of production of fissile materials made following the first Soviet A-bomb

[12]So named after Lawrence's death in 1958.

test and the sudden onset of the Korean War, and (4) the introduction of thermonuclear techniques, including variations on both the super and boosting.

The establishment of the second nuclear weapons laboratory at Livermore in effect doubled the number of scientists, engineers, and technicians working on nuclear weapons. From the beginning, the Livermore scientists probed new designs for small nuclear weapons as well as extensions of the Teller–Ulam design for very large weapons. But this doubling of the number of persons involved did much more than simply increase the rate of progress by a factor of two. As Teller had said in explaining why he left Los Alamos and undertook the promotion of a second laboratory, technology, like so many other elements of human endeavor, does thrive on competition. I have little doubt that the Livermore laboratory was seen as competition by Los Alamos, and I know for a fact that we at Livermore saw ourselves as being in competition with them.

Our first two test explosions at Nevada were programmed to have low yields but they actually resulted in even less (one of them barely destroyed the cab on top of its tower, leaving most of the tower still standing for all to see and poke fun at). On top of that, our first attempt to produce a large explosion at Bikini in the spring of 1954 resulted in a fizzle (the Koon shot; see the table in chapter 5), and we canceled plans for firing a second shot after we understood what went wrong and deduced it would only happen again without a substantial redesign and modification[18]. This poor beginning may have relieved some pressure at Los Alamos, but it stimulated us to gird our loins and do better next time.

The net result was that over the next decade, the two laboratories together produced a very great variety of designs suitable for a wide variety of purposes and fitted to many different kinds of delivery systems. These included warheads for ground-to-air missiles for continental air defense, such as the Nike Ajax, the Nike Hercules, and the Bomarc. The first was not actually deployed with a nuclear warhead, being superseded by the second before that could happen. Also included were warheads for tactical missiles such as the Corporal, the Pershing, and Redstone; nuclear shells for specially designed nuclear artillery pieces, atomic demolition devices to be used like gigantic mines for slowing an advance; a warhead small enough to be fitted to a bazooka-like device called the Davy Crockett; and all sorts of bombs for use on both land-based and carrier-based short range aircraft. Also included were nuclear explosives designed and sized for naval applications as mines, and as warheads for torpedos, and antisubmarine missiles such as ASROC.

Within ten years, the new burst of intellectual energy focused on nuclear weapons design problems resulted in bombs that weighed less than one hundred pounds (as the Davy Crockett), that were only inches in diameter (as for artillery shells), and that were on the average sufficiently efficient in their use of nuclear materials so that they could number many tens of thousands. Of this huge total, Secretary of Defense McNamara in the early 1960's reported that 7000 were located in Western Europe alone, and supposedly dedicated to the defense of that region.

The introduction of such a large variety of weapons led directly to their being stored in a great many different locations abroad as well as at home, and to a correspondingly great proliferation in the number of military command elements having immediate control of the individual weapons. There can be little doubt that the great leap forward in nuclear weapons design and deployment that occurred in the 1950's resulted from the factors listed at the beginning of this section. Whether this accelerated rate of progress has had a beneficial or harmful effect on national security is another interesting and important question, but it is beyond the scope of this book.[13]

The Oppenheimer Case

The primary dispute over the hydrogen bomb and the secondary arguments deriving from it—arguments over allocation of scarce resources, including personnel, money, and fissile materials—had

[13]In other books and articles I have given the reasons why I think it has been harmful.

from the beginning some underlying personal elements. Oppenheimer[19] at times could be a very arrogant man, and those with whom he argued these matters had from time to time been deeply wounded by his sharp wit and acerbic tongue Therefore it was perhaps only natural that very soon some of those with the more extreme views about the need for bigger bombs and other more exotic weapons systems began to question the motives of their chief opponent.

Much of the argument took place during the height of the McCarthy era. As a result, the fears that fed McCarthyism and the fears that McCarthyism engendered were both an essential part of the background against which the Oppenheimer drama was played. Beginning in the mid-thirties, and ending only well after he had become director of the Los Alamos laboratory, Oppenheimer had had a number of real, though superficial, associations with Communists and Communist causes. These were known to the various official bodies investigating such matters, and he himself several times appeared before the House Un-American Activities Committee (HUAC) as a cooperative witness in hearings focused on alleged "Red" political activities and associations on the part of others. McCarthy's Senate investigating committee in 1953 apparently considered taking up the matter of Oppenheimer's own Communist associations but is said to have been dissuaded from doing so by Richard Nixon, then the newly elected Vice President of the United States and formerly a member of HUAC[20].

Several of those most strongly opposed to Oppenheimer's views either already knew quite a bit about Oppenheimer's personal history, or were in situations where they were able to gain access to files containing detailed information about his past. Those in the former category included Lawrence and Teller and those in the latter category included William Borden on the JCAE staff and David Griggs and others in the air force. More importantly, Lewis Strauss and K. D. Nichols, respectively the newly appointed (in 1953) chairman and general manager of the AEC, fell into both categories. They had long known well Oppenheimer and his political idiosyncrasies. They had very often opposed him in public and private arguments and had as a consequence felt the combined force of his powerful intellect and his often arrogant attitude. And they had complete access to all records concerning him.

The air force took the first actions against Oppenheimer. Even as early as 1949, General Vandenberg, then air force chief of staff, is said to have insisted that Vannevar Bush rather than Oppenheimer serve as chairman of the special committee to evaluate the evidence of the first Russian A-bomb[21]. Later, in 1951, Secretary Finletter and General Vandenberg gave direct orders to Ivan Getting and Louis Ridenour, then the two top civilian scientists in air force headquarters, not to use Oppenheimer as a consultant in any further studies and to keep classified air force information away from him. These orders were issued when Getting and Ridenour proposed to set up a special study of strategic nuclear weapons under Oppenheimer's chairmanship analogous to those already under way for air defense (Project Charles) and tactical weaponry (Project Vista)[22].

The Department of Defense handled the matter very smoothly. In accordance with Reorganization Plan No. 6 (1953), the Research and Development Board (RDB) and all its subcommittees and panels were abolished and replaced by new structures. Oppenheimer's "need to know" as far as Office of the Secretary of Defense was concerned, had been based on his membership in one of the RDB's regular subcommittees (the Committee on Atomic Energy) and in the RDB's *ad hoc* Long Range Objectives Panel, and so by simply eliminating these and not appointing him as a consultant anywhere in the new structure, his clearance in that office was automatically voided. Defense Secretary Charles E. Wilson, in a press conference on April 14, 1954, said in reference to the Oppenheimer affair: "We dropped the whole board. That was a real smooth way of doing that one as far as the Defense Department was concerned."

The main blow came in the AEC in the spring of 1954, well after his direct influence over the affairs of that agency had virtually ceased. Back in June 1952, Oppenheimer offered Gordon Dean, the AEC chairman, his resignation from the GAC. He had been its chairman for five years and he seemed to feel that a change was timely; in addition he was well aware that he was increasingly out of step with the rest of the government and that as a result, his effectiveness had greatly diminished. Dean accepted Oppenheimer's resignation and his term came to an end on August 8, 1952. His status thereupon changed to that of consultant. That status kept his security clearance alive even though he was very little used in that capacity. During his long service on the GAC, Oppenheimer's classified files had been maintained under guard at the Institute for Advanced Study at Princeton, New Jersey. In December 1952, Dean ordered all those classified papers pertaining specifically to GAC affairs returned to Washington.

One of Dean's last acts before stepping down as chairman of the AEC on June 30, 1953, was to extend Oppenheimer's consultant contract for another year. On July 2, 1953, Lewis Strauss became chairman of the AEC. On July 7, as virtually his first act of business upon taking over the chairmanship, he ordered the removal of all remaining classified material from Oppenheimer's Princeton office. Oppenheimer remained an AEC consultant, but continued to do very little in that capacity.

On November 7, 1953, William Borden, who had recently left his position as executive director of the staff of the JCAE, sent a letter to J. Edgar Hoover, director of the Federal Bureau of Investigation (FBI), which said in part:

> The purpose of this letter is to state my own exhaustively considered opinion, based upon years of study of the available classified evidence, that more probably than not J. Robert Oppenheimer is an agent of the Soviet Union[23].

This letter and the other factors related above led to the complete cancellation of Oppenheimer's clearance, to an order by President Eisenhower to place a blank wall between Oppenheimer and the nation's secrets, and to formal hearings conducted by the AEC in April and May 1954 to determine whether or not he was a security risk.

Borden's conclusions were, as he explained in his letter, based on a multiplicity of allegedly interrelated matters. These included Oppenheimer's long associations with Communists and Communist causes and several instances of equivocal behavior related to safeguarding our vital secrets, especially the famous case involving

Haakon Chevalier. In addition, Borden also cited Oppenheimer's advice concerning the superbomb, the second laboratory and other related matters, all of which, in Borden's view, were deliberately designed to promote the best interests of the Soviets. On the other hand, when the AEC authorities converted these and other similar allegations into formal charges, they recognized the latter portion of Borden's accusation would be tantamount to trying Oppenheimer for his opinions and so they deliberately played them down[24]. That part of the AEC's charges that dealt with the superbomb controversy was therefore cast in terms designed to raise questions about Oppenheimer's veracity rather than in terms of how his opinions on programs affected national security. Accordingly, a December 23, 1953, letter from AEC General Manager K. D. Nichols to Oppenheimer setting forth the AEC's formal charges contains a very long list of Communist associations and allegedly improper and careless actions related to security matters, and only a very short section—about one-tenth as long—concerning his purportedly negative views on the superbomb and containing the allegation that he tried to dissuade "other outstanding scientists" from working on the project.

Most of the questions and most of the time during the hearings, however, had the opposite emphasis. They were devoted mainly to the matter of his advice rather than to his alleged Communist connections. The witnesses included several past and current senior government officials, several experts on questions of personnel security, and many of the scientists who had played a leading role in the wartime Manhattan project and the current hydrogen bomb project. A large majority testified in Oppenheimer's favor. Among this group were Oppenheimer's wartime boss, General Groves, and the two previous chairmen of the AEC, David Lilienthal and Gordon Dean. Not surprisingly, all of the scientists who agreed generally with his views on the superbomb strongly supported him. In addition several of those who had disagreed with him in the hydrogen bomb issue also supported him. These included John Von Neumann, Norris Bradbury, and Karl Compton. Those whose testimony was largely against Oppenheimer, or, in any event, those whose testimony had that effect, all were persons who had held strongly differing opinions on the urgency of the need for a super weapon and on the best means for developing it. This group included Luis Alvarez, Wendell Latimer, David Griggs, and Edward Teller. (Ernest Lawrence was ill at the time, or he would have been in this group.) Teller's views on what the important issues really were are fairly clear. When he was asked whether he considered Oppenheimer a security risk, he said,

> In a great number of cases I have seen Dr. Oppenheimer act—I understood that Dr. Oppenheimer acted—in a way which for me was exceedingly hard to understand. I thoroughly disagreed with him in numerous issues and his actions frankly appeared to me confused and complicated. To this extent I feel that I would like to see the vital interests of this country in hands which I understand better, and therefore trust more.
>
> In this very limited sense I would like to express a feeling that I would feel personally more secure if public matters would rest in other hands[25].

and again at another point in his testimony:

> I believe, and that is merely a question of belief and there is no expertness, no real information behind it, that Dr. Oppenheimer's character is such that he would not knowingly and willingly do anything that is designed to endanger the safety of this country. To the extent, therefore, that your question is directed toward intent, I would say I do not see any reason to deny clearance.
>
> If it is a question of wisdom and judgment, as demonstrated by actions since 1945, then I would say one would be wiser not to grant clearance. I must say that I am myself a little bit confused on this issue, particularly as it refers to a person of Oppenheimer's prestige and influence. May I limit myself to these comments?[26].

The hearing board voted two to one against Oppenheimer. Ward Evans, the only scientist on the board, and "a man of intractable political conservatism[27]," made up the minority. Gordon Gray, a man with considerable experience as a high government administrator, and Thomas Morgan, an industrial and financial executive, made up the majority.

K. D. Nichols, who as general manager had the formal responsibility in the matter, accepted the majority view, and the commission in turn backed him up by four to one. Again the minority of one consisted of the lone scientific member, Henry Smyth, author of the famous "Smyth report" [28] on the Manhattan project. The majority consisted of AEC Chairman Lewis Strauss plus Commissioners Joseph Campbell, Thomas Murray, and Eugene Zuckert.

As a final result, all of Oppenheimer's remaining government connections were severed, and his long and intensive service to national security came to an end. He continued to serve as director of the Institute of Advanced Study at Princeton, and it adds another note of irony to the whole situation to recall that Lewis Strauss was a trustee of the institute.

Epilogue—1975

In the years immediately following the finding that he was a security risk, neither Oppenheimer nor his friends ever pressed very hard for a retrial or any other formal sort of vindication. Even so, almost ten years later, on December 3, 1963, President Lyndon Johnson presented him with the Fermi Award, the highest honor the U.S. government can bestow for notable service in the field of nuclear energy. The plan for giving him this award was actually approved by John F. Kennedy shortly before his assassination. It was, of course, based in part on the importunings of Oppenheimer's friends and colleagues and in part on a visceral recognition by many, including no doubt Kennedy and perhaps also Johnson, that a great injustice had been done. There had been some residual opposition among Oppenheimer's old enemies, but this was overcome, in part because many of this last group had wanted for some time to give the Fermi Award to Teller, and it was obvious that that could not be done without giving it to Oppenheimer also. Teller, in fact, received the Fermi Award the previous year with tacit understanding by those concerned that Oppenheimer would then get it next. Oppenheimer died a few years later on February 18, 1967.

Edward Teller's main base of operations has remained at the Lawrence Livermore Laboratory ever since its founding in 1952. He served briefly as the laboratory director (1958–1960), but his considerable influence over the laboratory's life and scientific programs has always derived more from his scientific "charisma" than from his formal positions. However, ever since he led the winning side in the fight over the H-bomb, his main influence over the course of American nuclear affairs has been primarily of a political nature. Here also this influence has been exercised not so much through formal advisory channels (though he has in fact been a member of more official advisory bodies than Oppenheimer ever heard of) but rather more through direct and personal relationships with a host of high-level decision makers in the American government and business establishment. And not only has Teller been a key figure in the promotion of the uses of nuclear technology, but also, beginning with his fight against the nuclear test moratorium of 1958–1961, he has been the leading personality in opposing the various attempts to contain the nuclear arms race. In my opinion, he deserves very much of the credit (or blame), probably more than any other single individual, for the failure of the 1963 Nuclear Test Ban Treaty to prohibit underground tests along with those in all other environments, and for the inclusion in the 1968 Non-Proliferation Treaty of the provision making a special place for the so-called peaceful uses of nuclear explosives, a provision which in the long run will probably prove the undoing of that treaty if something else does not do so first.

In addition to Robert Oppenheimer, several others among the key scientific figures in this history died soon after the events described here had worked their course. Enrico Fermi died in 1954, at age fifty-three while still a very active scientist and teacher at the University of Chicago. John Von Neumann died in 1957 at age fifty-three after a very brief but spectacularly influential career as a top-level advisor in the Defense Department and as a member of the AEC. Ernest Lawrence died in 1958, at age fifty-six, as the result of a flare-up in his underlying chronic illness directly associated with his participation in the 1958 Geneva Conference on the Test Ban.

Glenn Seaborg, the one GAC member who was absent at the time of the October 1949 meeting, continued a productive career in nuclear chemistry at the Lawrence Berkeley Laboratory and was awarded the Nobel Prize in chemistry in 1951 for his accomplishments in that field. In 1961, President Kennedy appointed him chairman of the AEC, and he continued in that post through the Johnson years and into the beginning of the Nixon administration. He is now once again a professor at Berkeley.

Luis Alvarez similarly continued a very productive career in nuclear physics at the Lawrence Berkeley Laboratory, and was awarded the Nobel Prize in physics in 1968 for his many accomplishments in that field. Although he has continued to serve occasionally on high-level advisory bodies, he never again played such a politically important role as that described in this book.

I. I. Rabi, the other signer of the minority addendum, for many years after continued a mixed career as an academic physicist at Columbia University and as a high-level government advisor. Most interesting from the point of view of this history, he succeeded Robert Oppenheimer as chairman of the GAC following the latter's resignation from this post in 1952, and hence he presided over that committee during the period of the Oppenheimer hearings. In addition, during the fifties he was first a member and then chairman of the Science Advisory Committee to the Office of Defense Mobilization, and then a member of its successor, the President's Science Advisory Committee. During the sixties, he was also a charter member of the General Advisory Committee on Arms Control and Disarmament. He is now retired from most such activities and lives in New York City.

Norris Bradbury continued to serve as director of the Los Alamos Scientific Laboratory until his retirement in 1970. The laboratory prospered throughout that entire period. He always promoted the development of nuclear weapons and other military (and civilian) applications of nuclear technology, but he also was tolerant of the idea that an appropriately safeguarded test ban (and other arms control and disarmament measures) could be in the best interests of U.S. national security.

Stan Ulam and Carson Mark have continued to work on nuclear weapon design problems to the present time. Mark remained as head of the Los Alamos Theoretical Division until early 1974, at which time he retired from that post, but he still lives at Los Alamos and serves as a consultant to the laboratory. Ulam left full-time employment at Los Alamos not long after he made his famous contribution, and reentered academic life, mainly at the University of Colorado. He has, however, continued throughout these last twenty years to serve as a consultant to Los Alamos and to contribute his ideas to various of the laboratory's programs.

Hans Bethe has just retired after almost forty years of service as Professor of Physics at Cornell University. He received the Nobel Prize in 1967 for the pioneering work he did in the 1930's on thermonuclear processes in stars and on other fundamental nuclear problems. Since the events reported here, he has continued to participate in applied nuclear science and in nuclear politics in a quiet but effective way as a member of several important advisory committees, including the President's Science Advisory Committee; as a consultant to Los Alamos, the General Atomics Corporation, the Avco Corporation, and other high technology institutions; and by occasionally writing and speaking on these subjects.

Among the younger scientists mentioned in this chronicle, those with the most remarkable subsequent careers are Harold Brown, John Foster, Richard Garwin, Fred Reines, Theodore Taylor and Frederick DeHoffmann. Brown subsequently served on important advisory committees including the President's Science Advisory Committee, and in high government posts, including the director of Defense Research and Engineering (1961–1965) and secretary of the air force (1965–1969). He is now president of the California Institute of Technology. In addition, he was a charter member of the U.S. delegation to SALT beginning in 1969. Foster also served as director of Defense Research and Engineering for an exceptionally long time (1965–1973). He is now a vice president of TRW, responsible for that organization's broad attack on the energy problem. Richard Garwin joined International Business Machines (IBM) soon after the events described in this book, and has had a very productive career there ever since. More importantly, he has devoted a very large part of his time and energy to service on various public bodies, including the President's Science Advisory Committee, and he has worked on and effectively influenced a large number of special public policy issues having a high technological content, including national security matters, arms control, space and the supersonic transport (SST). Reines later went into a strictly academic career. He made the first successful observation of the elusive neutrino, and has been a dean at the University of California, Irvine, where he currently works as a Professor of Physics. Taylor acquired considerable reknown in the seventies as a result of his studies of and warnings about the possibilities of theft of fissile material by nongovernment groups and its subsequent use for all sorts of dreadful purposes. In the late fifties DeHoffmann became founding president of General Atomics in La Jolla, and in the seventies he became the chief executive of the Salk Institute for Biological Studies next door.

Many of the political figures importantly involved in this history never again played major roles in succeeding events. David Lilienthal retired from the scene immediately after Truman's decision, Senator McMahon died that same spring, and Louis Johnson, Robert LeBaron, and Dean Acheson had left the government well before the end of this particular history in 1954.

Lewis Strauss continued to serve as chairman of the AEC for several years after the matter of Oppenheimer's fitness to serve as an advisor was formally decided. Then in 1958, Eisenhower named him Secretary of Commerce on an interim basis—Congress was not then in session—and placed his name before the Senate as his nominee for a regular appointment in that post. In one of history's more remarkable ironies, his nomination was rejected, largely as a result of personal animosities arising from his prior service on the AEC. Strauss never again played an important role in American politics. He died in the spring of 1974, at age seventy-seven.

Igor Kurchatov continued as the technical director of the Soviet nuclear weapons program until his death in 1960 at age fifty-seven. During the late fities it is reported that he played an important part in persuading Soviet authorities, particularly Khrushchev, to move towards a nuclear test ban. During his life, he was much admired by his co-workers and aides, and after his death, he became something of a modern popular hero, Soviet style. Two books and several long articles have been written in this vein about him. These, together with two speeches he made at Party Congresses, and one in England, provide much of what is known about the Soviet program of those days. His main laboratory, in Moscow, is now known as the Kurchatov Institute and today houses a major part of the Soviet work designed to produce thermonuclear energy under controlled (that is, nonexplosive) conditions.

Several of the other Russians mentioned in this chronicle are still active in nuclear affairs. Igor Golovin works in the controlled thermonuclear program at the Kurchatov Institute, Peter Kapitza directs work on the same problem at his own institute, A. P. Vinovgradov remains active in academy affairs, and Vassily Emelyanov still works on the international aspects of nuclear energy, including questions of nuclear arms control. The last three have also been active for some years in the international Pugwash movement, which has stopping and then reversing the nuclear arms race as its main objective. Andrei Sakharov continued to work on the design of nuclear weapons until sometime in the early sixties when he finally turned away from such work as a matter of conscience. He still lives in Russia and continues to work on theoretical physics problems. He is best known in the West as one of the leading dissidents on the Moscow scene. He participates in a wide range of political activities that appear to have stemmed originally from his early (1958–1959), and evidently impassioned, efforts to stop nuclear fallout in particular and the nuclear arms race in general.

APPENDIX

The GAC Report of October 30, 1949

GENERAL ADVISORY COMMITTEE
to the
U.S. ATOMIC ENERGY COMMISSION
Washington 25, D.C.

October 30, 1949

Dear Mr. Lilienthal:

At the request of the Commission, the seventeenth meeting of the General Advisory Committee was held in Washington on October 29 and 30, 1949 to consider some aspects of the question of whether the Commission was making all appropriate progress in assuring the common defense and security. Dr. Seaborg's absence in Europe prevented his attending this meeting. For purposes of background, the Committee met with the Counsellor of the State Department, with Dr. Henderson of AEC Intelligence, with the Chairman of the Joint Chiefs of Staff, the Chairman of the Military Liaison Committee, the Chairman of the Weapons Systems Evaluation Group, General Norstadt and Admiral Parsons. In addition, as you know, we have had intimate consultations with the Commission itself.

The report which follows falls into three parts. The first describes certain recommendations for action by the Commission directed toward the common defense and security. The second is an account of the nature of the super project and of the super as a weapon, together with certain comments on which the Committee is unanimously agreed. Attached to the report, but not a part of it, are recommendations with regard to action on the super project which reflect the opinions of Committee members.

The Committee plans to hold its eighteenth meeting in the city of Washington on December 1, 2 and 3, 1949. At that time we hope to return to many of the questions which we could not deal with at this meeting.

J. R. Oppenheimer
Chairman

(continued)

UNITED STATES ATOMIC ENERGY COMMISSION
WASHINGTON, D.C. 20545
HISTORICAL DOCUMENT NUMBER 349

David E. Lilienthal
Chairman
U.S. Atomic Energy Commission
Washington 25, D.C.

PART I

(1) PRODUCTION. With regard to the present scale of production of fissionable material, the General Advisory Committee has a recommendation to make to the Commission. We are not satisfied that the present scale represents either the maximum or the optimum scale. We recognize the statutory and appropriate role of the National Military Establishment in helping to determine that. We believe, however, that before this issue can be settled, it will be desirable to have from the Commission a careful analysis of what the capacities are which are not now being employed. Thus we have in mind that an acceleration of the program on beneficiation of low grade ores could well turn out to be possible. We have in mind that further plants, both separation and reactor, might be built, more rapidly to convert raw material into fissionable material. It would seem that some notion of the costs, yields and time scales for such undertakings would have to precede any realistic evaluation of what we should do. We recommend that the Commission undertake such studies at high priority. We further recommend that projects should not be dismissed because they are expensive but that their expense be estimated.

(2) TACTICAL DELIVERY. The General Advisory Committee recommends to the Commission an intensification of efforts to make atomic weapons available for tactical purposes, and to give attention to the problem of integration of bomb and carrier design in this field.

(3) NEUTRON PRODUCTION. The General Advisory Committee recommends to the Commission the prompt initiation of a project for the production of freely absorbable neutrons. With regard to the scale of this project the figure * * * per day may give a reasonable notion. Unless obstacles appear, we suggest that the expediting of design be assigned to the Argonne National Laboratory.

With regard to the purposes for which these neutrons may be required, we need to make more explicit statements. The principal purposes are the following:

(a) The production of U-233.
(b) The production of radiological warfare agents.
(c) Supplemental facilities for the test of reactor components.
(d) The conversion of U-235 to plutonium.
(e) A secondary facility for plutonium production.
(f) The production of tritium (1) for boosters, (2) for super bombs.

We view these varied objectives in a quite different light. We have a great interest in the U-233 program both for military and for civil purposes. We strongly favor, subject to favorable outcome of the 1951 Eniwetok tests, the booster program. With regard to radiological warfare, we would not wish to alter the position previously taken by our Committee. With regard to the conversion to plutonium, we would hardly believe that this alone could justify the construction of these reactors, though it may be important should unanticipated difficulties appear in the U-233 and booster programs. With regard to the use of tritium in the super bomb, it is our unanimous hope that this will not prove necessary. It is the opinion of the majority that the super program itself should not be undertaken and that the Commission and its contractors understand that construction of neutron producing reactors is not intended as a step in the super program.

PART II

SUPER BOMBS

The General Advisory Committee has considered at great length the question of whether to pursue with high priority the development of the super bomb. No member of the Committee was willing to endorse this proposal. The reasons for our views leading to this conclusion stem in large part from the technical nature of the super and of the work necessary to establish it as a weapon. We therefore here transmit *an elementary* account of these matters.

The basic principle of design of the super bomb is the ignition of the thermo-nuclear DD reaction by the use of a fission bomb, and of high temperatures, pressure, and neutron densities which accompany it. In overwhelming probability, tritium is required as an intermediary, more easily ignited than the deuterium itself and, in turn, capable of igniting the deuterium. The steps which need to be taken if the super bomb is to become a reality include:

(1) The provision of tritium in amounts perhaps of several °°°°° per unit.
(2) Further theoretical studies and criticisms aimed at reducing the very great uncertainties still inherent in the behavior of this weapon under extreme conditions of temperature, pressure and flow.
(3) The engineering of designs which may on theoretical grounds appear hopeful, particularly with regard to the °°°°° problems presented.
(4) Carefully instrumented test programs to determine whether the deuterium-tritium mixture will be ignited by the fission bomb, °°°°°.

It is notable that there appears to be no experimental approach short of actual test which will substantially add to our conviction that a given model will or will not work, and it is also notable that because of the unsymmetric and extremely unfamiliar conditions obtaining, some considerable doubt will surely remain as to the soundness of theoretical anticipation. Thus we are faced with a development which cannot be carried to the point of conviction without the actual construction and demonstration of the essential elements of the weapon in question. This does not mean that further theoretical studies would be without avail. It does mean that they could not be decisive. A final point that needs to be stressed is that many tests may be required before a workable model has been evolved or before it has been established beyond reasonable doubt that no such model can be evolved. Although we are not able to give a specific probability rating for any given model, we believe that an imaginative and concerted attack on the problem has a better than even chance of producing the weapon within five years.

A second characteristic of the super bomb is that once the problem of initiation has been solved, there is no limit to the explosive power of the bomb itself except that imposed by requirements of delivery. This is because one can continue to add deuterium—an essentially cheap material—to make larger and larger explosions, the energy release and radioactive products of which are both proportional to the amount of deuterium itself. Taking into account the probable limitations of carriers likely to be available for the delivery of such a weapon, it has generally been estimated that the weapon would have an explosive effect some hundreds of times that of present fission bombs. This would correspond to a damage area of the order of hundreds of square miles, to thermal radiation effects extending over a comparable area, and to very grave contamination problems which can easily be made more acute, and may possibly be rendered less acute, by surrounding the deuterium with uranium or other material. It needs to be borne in mind that for delivery by ship, submarine or other such carrier, the limitations here outlined no longer apply and that the weapon is from a technical point of view without limitations with regard to the damage that it can inflict.

It is clear that the use of this weapon would bring about the destruction of innumerable human lives; it is not a weapon which can be used exclusively for the destruction of material installations of military or semi-military purposes. Its use therefore carries much further than the atomic bomb itself the policy of exterminating civilian populations. It is of course true that super bombs which are not as big as those here contemplated could be made, provided the initiating mechanism works. In this case, however, there appears to be no chance of their being an economical alternative to the fission weapons themselves. It is clearly impossible with the vagueness of design and the uncertainty as to performance as we have them at present to give anything like a cost estimate of the super. If one uses the strict criteria of damage area per dollar and if one accepts the limitations on air carrier capacity likely to obtain in the years immediately ahead, it appears uncertain to us whether the super will be cheaper or more expensive than the fission bomb.

PART III

Although the members of the Advisory Committee are not unanimous in their proposals as to what should be done with regard to the super bomb, there are certain elements of unanimity among us. We all hope that by one means or another, the development of these weapons can be avoided. We are all reluctant to see the United States take the initiative in precipitating this development. We are all agreed that it would be wrong at the present moment to commit ourselves to an all-out effort toward its development.

We are somewhat divided as to the nature of the commitment not to develop the weapon. The majority feel that this should be an unqualified commitment. Others feel that it should be made conditional on the response of the Soviet government to a proposal to renounce such development. The Committee recommends that enough be declassified about the super bomb so that a public statement of policy can be made at this time. Such a statement might in our opinion point to the use of deuterium as the principal source of energy. It need not discuss initiating mechanisms nor the role which we believe tritium will play. It should explain that the weapon cannot be explored without developing it and proof-firing it. In one form or another, the statement should express our desire not to make this development. It should explain the scale and general nature of the destruction which its use would entail. It should make clear that there are no known or foreseen nonmilitary applications of this development. The separate views of the members of the Committee are attached to this report for your use.

J. R. Oppenheimer

■ ■ ■

October 30, 1949

We have been asked by the Commission whether or not they should immediately initiate an "all-out" effort to develop a weapon whose energy release is 100 to 1000 times greater and whose destructive power in terms of area of damage is 20 to 100 times greater than those of the present atomic bomb. We recommend strongly against such action.

We base our recommendation on our belief that the extreme dangers to mankind inherent in the proposal wholly outweigh any military advantage that could come from this development. Let it be clearly realized that this is a super weapon; it is in a totally different category from an atomic bomb. The reason for developing such super bombs would be to have the capacity to devastate a vast area with a single bomb. Its use would involve a decision to slaughter a vast number of civilians. We are alarmed as to the possible global effects of the radioactivity generated by the explosion of a few super bombs of conceivable magnitude. If super bombs will work at all, there is no inherent limit in the destructive power that may be attained with them. Therefore, a super bomb might become a weapon of genocide.

The existence of such a weapon in our armory would have far-reaching effects on world opinion: reasonable people the world over would realize that the existence of a weapon of this type whose power of destruction is essentially unlimited represents a threat to the future of the human race which is intolerable. Thus we believe that the psychological effect of the weapon in our hands would be adverse to our interest.

We believe a super bomb should never be produced. Mankind would be far better off not to have a demonstration of the feasibility of such a weapon until the present climate of world opinion changes.

It is by no means certain that the weapon can be developed at all and by no means certain that the Russians will produce one within a decade. To the argument that the Russians may succeed in developing this weapon, we would reply that our undertaking it will not prove a deterrent to them. Should they use the weapon against us, reprisals by our large stock of atomic bombs would be comparably effective to the use of a super.

In determining not to proceed to develop the super bomb, we see a unique opportunity of providing by example some limitations on the totality of war and thus of limiting the fear and arousing the hopes of mankind.

James B. Conant
Hartley Rowe
Cyril Stanley Smith
L. A. DuBridge
Oliver E. Buckley
J. R. Oppenheimer

(continued)

■ ■ ■

October 30, 1949

AN OPINION ON THE DEVELOPMENT OF THE "SUPER"

A decision on the proposal that an all-out effort be undertaken for the development of the "Super" cannot in our opinion be separated from considerations of broad national policy. A weapon like the "Super" is only an advantage when its energy release is from 100–1000 times greater than that of ordinary atomic bombs. The area of destruction therefore would run from 150 to approximately 1000 square miles or more.

Necessarily such a weapon goes far beyond any military objective and enters the range of very great natural catastrophes. By its very nature it cannot be confined to a military objective but becomes a weapon which in practical effect is almost one of genocide.

It is clear that the use of such a weapon cannot be justified on any ethical ground which gives a human being a certain individuality and dignity even if he happens to be a resident of an enemy country. It is evident to us that this would be the view of peoples in other countries. Its use would put the United States in a bad moral position relative to the peoples of the world.

Any postwar situation resulting from such a weapon would leave unresolvable enmities for generations. A desirable peace cannot come from such an inhuman application of force. The postwar problems would dwarf the problems which confront us at present.

The application of this weapon with the consequent great release of radioactivity would have results unforeseeable at present, but would certainly render large areas unfit for habitation for long periods of time.

The fact that no limits exist to the destructiveness of this weapon makes its very existence and the knowledge of its construction a danger to humanity as a whole. It is necessarily an evil thing considered in any light.

For these reasons we believe it important for the President of the United States to tell the American public, and the world, that we think it wrong on fundamental ethical principles to initiate a program of development of such a weapon. At the same time it would be appropriate to invite the nations of the world to join us in a solemn pledge not to proceed in the development or construction of weapons of this category. If such a pledge were accepted even without control machinery, it appears highly probable that an advanced stage of development leading to a test by another power could be detected by available physical means. Furthermore, we have in our possession, in our stockpile of atomic bombs, the means for adequate "military" retaliation for the production or use of a "super."

E. Fermi
I. I. Rabi

Notes

Notes for Chapter ONE

1. Robert Gilpin, *American Scientists and Nuclear Weapons Policy*, Princeton, N.J.: Princeton University Press, 1962.

Notes for Chapter TWO

1. The creation, operation, and accomplishments of the Manhattan project are described in: Henry D. Smyth, *Atomic Energy for Military Purposes*, Princeton, N.J.: Princeton University Press, 1945; R. Hewlett and O. F. Anderson, Jr., *The New World*, University Park, Pa.: Pennsylvania State University Press, 1962.
2. The number twelve appears in Edward Teller with Allen Brown, *The Legacy of Hiroshima*, Garden City, N.Y.: Doubleday & Company, 1962, and in a press conference held by Bradbury on September 24, 1954.
3. Press release, Los Alamos Scientific Laboratory, September 24, 1954.
4. Richard G. Hewlett and F. Duncan, *A History of the U.S. Atomic Energy Commission*, Vol II, *The Atomic Shield*, University Park, Pa.:

Pennsylvania State University Press, 1969, p. 178 (hereafter cited as *USAEC History*).

5. Edward Teller, "The Work of Many People," *Science,* February 25, 1955, p. 268.
6. LA-575 Los Alamos Scientific Laboratory, University of California, unclassified edited version released May 17, 1971.
7. Hewlett and Duncan, *USAEC History,* p. 59.
8. Letter, Robert Oppenheimer to James Conant, October 1949, in Headquarters Records, USAEC, Washington, D.C. Reproduced in part in J. R. Shepley and C. Blair, Jr., *The Hydrogen Bomb,* New York: Greenwood, 1954, p. 70.
9. I recall his saying this at a meeting of Serber, Lawrence, and myself at Berkeley in the spring of 1950.
10. Teller with Brown, *The Legacy of Hiroshima,* p. 50.
11. USAEC, *In the Matter of J. Robert Oppenheimer,* Cambridge, Mass.: The M.I.T. Press, 1971, p. 487.

Notes for Chapter THREE

1. There are now two good sources of information on the Soviet program which nicely complement each other. One is I. N. Golovin, *I. V. Kurchatov,* 2nd ed., Moscow: Atomizdat, 1973, and the other is Arnold Kramish, *Atomic Energy in the Soviet Union,* Stanford, Cal.: Stanford University Press, 1959. The first edition of Golovin's biography has been translated by William H. Dougherty, Bloomington, Ind.: Selbstverlag Press, 1968. Golovin is himself a physicist, and worked closely with Kurchatov from 1943 (or 1944) until the latter's death. Kramish's book contains only unclassified data, but Kramish spent some years at the RAND Corporation and may be presumed to have had access to other information as well. Other interesting sources are P. Astashenkov, *Kurchatov,* Moscow, 1967, and a pair of articles by V. C. Emelyanov in *Yunost,* 1968, No. 4, pp. 83–93, and No. 5, pp. 88–95, entitled "Kurchatov, As I Knew Him." All of these sources plus personal conversations were used in writing this section.
2. Private communication.
3. Golovin, *I. V. Kurchatov,* pp. 41, 48.
4. *Izvestia,* October 14, 1941, p. 3.
5. Harry S. Truman, *Memoirs,* Vol. 1, *Year of Decisions.*
6. *New York Times,* March 2, 1969, p. 7, col. 1.
7. Smyth, *Atomic Energy for Military Purposes.*

8. Private communication.
9. *I. V. Kurchatov*, p. 76. My translation. The first edition gives September 23 as the date of the explosion. That was, in fact, the date of Truman's announcement.
10. The remark about "those asiatics" was reported to me many years ago by a U.S. senator who was both a friend of Truman's and an expert on atomic energy affairs. The sentence about the signatures is based on a recent conversation with Bacher.
11. Dexter Masters and Katherine Way, eds., *One World or None*, New York: McGraw-Hill, 1946.
12. L. L. Strauss, *Men and Decisions*, Garden City, N.Y.: Doubleday & Company, 1962, p. 439. Bush made other longer estimates later on, apparently using differing assumptions.
13. *Science*, January 27, 1961, p. 261.
14. *I. V. Kurchatov* (Second edition only!), p. 59.
15. As reported in the *New York Times*, August 17, 1945, p. 4.
16. The full text of the Franck report can be found in M. Grodzins and E. Rabinowitch, eds., *The Atomic Age, Scientists in National and World Affairs*, New York: Basic Books, 1963, pp. 19–29.
17. Hadley Cantril, *Public Opinion 1935–1946*, Princeton, N.J.: Princeton University Press, 1957, pp. 22–23.
18. JCAE, *Soviet Atomic Espionage*, Government Printing Office, April 1951, pp. 5–6.
19. Kramish, *Atomic Energy in the Soviet Union*, pp. 109–110.
20. A. A. Zborskin et al., eds., *Biograficheskiy Slovar, deyateley estestvoznaniya i tekhniki*, Moscow: Gosudarstvennoye Nauchnoye Press, 1958.
21. For example, see Strauss, *Men and Decisions*, p. 218.

Notes for Chapter FOUR

1. *Washington Post*, November 18, 1949, p. 1.
2. *New York Times*, January 17, 1950, p. 1.
3. *Bulletin of the Atomic Scientists*, Vol. V, No. 10, October 1949, p. 265.
4. *Bulletin of the Atomic Scientists*, Vol. V. No. 11, November 1949, p. 294.
5. *Bulletin of the Atomic Scientists*, Vol. V, No. 10, October 1949, p. 275.
6. *Bulletin of the Atomic Scientists*, Vol. V, No. 11, November 1949, p. 323.

7. J. Robert Oppenheimer, "Physics in the Contemporary World," *Bulletin of the Atomic Scientists,* Vol. IV, No. 3, March 1948, p. 66.
8. Acting Chairman Sumner Pike *as quoted* in Hewlett and Duncan, *USAEC History*, p. 380.
9. Pike's request *as summarized* by the authors in *USAEC History,* p. 380.
10. Oppenheimer to Conant, October 12, 1949, in Headquarters Records, USAEC, Washington, D.C. Further quotations from this letter may be found in Hewlett and Duncan, *USAEC History,* and in Shepley and Blair, *The Hydrogen Bomb,* pp. 69–70.
11. Hewlett and Duncan, *USAEC History,* p. 396.
12. Where not otherwise indicated, I have used Hewlett and Duncan, *USAEC History,* chapter 12, as the source of the commissioners' views.
13. Hewlett and Duncan, *USAEC History,* p. 107. Also see David E. Lilienthal, *The Atomic Energy Years, 1945–1950,* New York: Harper & Row, 1964, pp. 623–632.
14. Strauss to Truman, November 25, 1949, in Headquarters Records, USAEC, Washington, D.C. The letter and its addendum are most readily available in Strauss, *Men and Decisions,* pp. 219–222, where they are published in full.
15. Hewlett and Duncan, *USAEC History,* p. 372. They give the original reference as JCAE, transcript of hearing, September 29, 1949, in AEC files.
16. Teller and Brown, *The Legacy of Hiroshima,* p. 44.
17. Hewlett and Duncan, *USAEC History,* p. 393.
18. *Ibid.,* p. 394.
19. *Ibid.,* p. 402.
20. From a speech made in Washington, February 2, 1950, according to the *New York Times,* February 3, 1950, p. 2.
21. Shepley and Blair, *The Hydrogen Bomb,* p. 88.
22. K. T. Compton to Truman, November 9, 1949, in Washington National Records Center, Modern Military Records Division, Suitland, Md. This letter is most readily available in Strauss' *Men and Decisions,* where it is reproduced in full on p. 440.
23. Hewlett and Duncan, *USAEC History,* p. 395.
24. USAEC, *In the Matter of J. Robert Oppenheimer,* the USAEC report of the 1954 Security Hearing, p. 87.
25. Edward Teller, "How Dangerous are Atomic Weapons?" *Bulletin of the Atomic Scientists,* Vol. III, No. 2, February 1947, pp. 35–36.

26. Hewlett and Duncan, *USAEC History*, p. 392.
27. Herbert Childs, *An American Genius*, New York: Dutton, 1968, p. 405.
28. Lilienthal, *The Atomic Energy Years.*
29. *Ibid.*, p. 577.
30. USAEC, *In the Matter of J. Robert Oppenheimer*, The M.I.T. Press, p. 774.
31. Lilienthal, *The Atomic Energy Years*, p. 614.
32. A speech at a University of Virginia alumni gathering, *New York Times*, February 3, 1950, p. 2.
33. Hewlett and Duncan, *USAEC History*, p. 403.
34. Dean Acheson, *Present at the Creation, My Years in the State Department*, New York: W. W. Norton & Company, 1969.
35. George F. Kennan, *Memoirs 1925–1950*, Boston: Little, Brown and Company, 1967, p. 471, *et seq.*
36. From Lilienthal's report on the meeting of the Special NSC Committee on the Superbomb, January 31, 1950, in Headquarters Records, USAEC, Washington, D.C. The material reproduced here may be most conveniently found in Lilienthal, *The Atomic Energy Years*, p. 624.
37. *Ibid*, p. 632.
38. Hewlett and Duncan, *USAEC History*, pp. 406–408. I have had further recent correspondence with Hewlett confirming this view.
39. *Foreign Relations Journal*, June 1969, p. 28.
40. Shepley and Blair, *The Hydrogen Bomb*. Shepley and Blair, were a pair of reporters who interviewed many of those most involved in these and the ensuing events within just a few years after they happened.
41. *New York Times*, February 1, 1950, p. 1.
42. These quotations are from the account given in Warner Schilling, "The H-Bomb Decision," *Political Science Quarterly*, Vol. 76, 1961, pp. 21–46.
43. A directive to Acheson and Johnson, as paraphrased in *USAEC History*, p. 528.
44. *New York Times*, February 1, 1950, p. 3.
45. *Ibid.*
46. *Bulletin of the Atomic Scientists*, Vol. VI, No. 2, February 1950, p. 71
47. *Bulletin of the Atomic Scientists*, Vol. VI, No. 3, March 1950, pp. 72–73.
48. *Bulletin of the Atomic Scientists*, Vol. VI, No. 7, July 1950, p. 75.

49. *Bulletin of the Atomic Scientists,* Vol. VI, No. 3, March 1950, and *Scientific American,* (March 1950), pp. 11–15; (April 1950), pp. 18–23; (May 1950), pp. 11–15.
50. See for instance Senator Jackson's colloquy in the hearings on the ABM (antiballistic missile) in 1970. Committee on Armed Service, USS, *Hearings on Authorizations for Military Procurement FY 1970,* Part 2, p. 1197.

Notes for Chapter FIVE

1. Teller, "The Work of Many People," *Science,* p. 272.
2. *Ibid,* p. 273.
3. S. M. Ulam, *Adventures of a Mathematician,* New York: Charles Scribner's Sons, forthcoming 1976. This quotation is from a manuscript provided by the author.
4. Press conference by Norris Bradbury, September 24, 1954, Los Alamos, N.Mex.
5. Teller, "The Work of Many People," p. 273.
6. Recent conversation with Edward Freeman.
7. USAEC, *In the Matter of J. Robert Oppenheimer,* The M.I.T. Press, p. 251.
8. Director of Military Application, USAEC, *Thermonuclear Research at the University of California Radiation Laboratory,* AEC 425/20, Washington, D.C., June 13, 1952.
9. John McPhee, "The Curve of the Blinding Energy Crisis," *New Yorker* magazine, December 2, 10, and 17, 1973.
10. The following passage and other related material is from Golovin, *I. V. Kurchatov,* chapter XVI; translation by this author.
11. Article by Alfred Friendly, *Washington Post,* November 18, 1949, p. 1. This article referred in turn to an earlier but largely unnoticed TV broadcast by Senator Edwin Johnson on November 1, 1949.
12. Andrei Sakharov, "How I Came to Dissent," *New York Review of Books,* March 21, 1974, pp. 11 ff.
13. As in Samuel Glasstone, ed., *The Effects of Nuclear Weapons,* Washington, D.C.: USAEC, 1962, p. 680.
14. *Pravda,* August 20, 1953, p. 2.
15. Glasstone, *op. cit.*
16. Speech at Bangalore, India, November 26, 1955, as reported in U.S.S.R. International Service, November 28, 1955; *Pravda,* November 27, 1955, p. 2 (TASS announcement); *Pravda,* November 28, 1955, p. 1 (Khrushchev speech). It was reported by others present that in the speech he actually said "a megaton" but all Soviet sources say a "few megatons" (neskolkikh millionov tonn).

Notes for Chapter SIX

1. According to a letter from Sterling Cole, chairman of the JCAE to Ernest Lawrence, November 17, 1953. In the E. O. Lawrence Archives, Bancroft Library.
2 USAEC, *In the Matter of J. Robert Oppenheimer,* The M.I.T. Press, p. 248.
3. Schilling, "The H-Bomb Decision," pp. 24–46. This is the best discussion of Truman's role in this matter that I am aware of. Other useful discussions can be found in Samuel P. Huntington, *The Common Defense,* New York: Columbia University Press, 1961; George Quester, *Nuclear Diplomacy, The First Twenty-Five Years,* New York: Dunellen Company; and in Gilpin, *American Scientists and Nuclear Weapons Policy.*
4. Schilling, "The H-Bomb Decision," p. 44.

Notes for Chapter SEVEN

1. USAEC, *In the Matter of J. Robert Oppenheimer,* The M.I.T. Press, p. 360.
2. A thorough discussion of NSC 68 can be found in Paul Y. Hammond, *NSC 68: Prologue to Disarmament,* in Warren R. Schilling et al., *Strategy, Politics and Defense Budgets.* New York: Columbia University Press, 1962.
3. Hewlett and Duncan, *USAEC History,* p. 531.
4. Alvarez testimony, USAEC, *In the Matter of J. Robert Oppenheimer,* The M.I.T. Press, p. 789.
5. J. Robert Oppenheimer, "Atomic Weapons and American Policy," *Foreign Affairs,* Vol. 31, No. 4, July 1953, p. 525.
6. USAEC, *In the Matter of J. Robert Oppenheimer,* The M.I.T. Press, p. 95.
7. Philip Stern, *The Oppenheimer Case,* New York, 1969, p. 172.
8. USAEC, *In the Matter of J. Robert Oppenheimer,* The M.I.T. Press, p. 684.
9. USAEC, *In the Matter of J. Robert Oppenheimer,* The M.I.T. Press, p. 749.
10. The only published account of the MTA project is a "gee whiz" article by Allen P. Armagnac, "The Most Fantastic Atom-Smasher," in *Popular Science,* November 1958, p. 108 *et seq.* References to the project can, of course, be found in Hewlett and Duncan, *USAEC History,* and in Childs, *An American Genius.*

11. USAEC, *In the Matter of J. Robert Oppenheimer,* The M.I.T. Press, pp. 81, 229, 251.
12. For instance, in Teller, *The Legacy of Hiroshima;* Len Giovannitti and Fred Freed, *The Decision to Drop the Bomb,* New York: Coward-McCann, 1965; Astashenkov, *Kurchatov;* and Golovin, *I. V. Kurchatov.*
13. See also Childs' account of this in Childs, *An American Genius.*
14. Teller, *The Legacy of Hiroshima,* pp. 54–55.
15. Hearing transcript, USAEC, *In the Matter of J. Robert Oppenheimer,* The M.I.T. Press, p. 248.
16. Director of Military Applications, USAEC, *Thermonuclear Research at the University of California Radiation Laboratory,* AEC 425/20, Washington, June 13, 1952.
17. Shepley and Blair, *The Hydrogen Bomb.* Excerpts were published in *Life* magazine.
18. Teller, *The Legacy of Hiroshima.*
19. The primary source for material on this subject is USAEC, *In the Matter of J. Robert Oppenheimer,* The M.I.T. Press. This edition consists of the hearing transcript, plus a foreword by Philip Stern, and the full text of several other relevant documents. Most important, it also has an index, which is lacking in the original Government Printing Office version (1954) and which makes the latter very hard to work with. The notoriety of the case and the circumstances surrounding it have given rise to much secondary literature. Of all this, I believe the best by far is Philip Stern's *The Oppenheimer Case.* Anyone wishing to understand the case better should also read Teller, *The Legacy of Hiroshima,* and Shepley and Blair, Jr., *The Hydrogen Bomb.*
20. Stern, *The Oppenheimer Case,* p. 204.
21. From a sworn statement by E. O. Lawrence, in USAEC, *In the Matter of J. Robert Oppenheimer,* The M.I.T. Press, p. 969, and testimony by Luis Alvarez, *ibid.,* p. 787. Not corroborated (but not really contradicted) by Vannevar Bush testimony, *ibid,* p. 910.
22. From recent correspondence and conversations with Ivan Getting.
23. USAEC, *In the Matter of J. Robert Oppenheimer,* The M.I.T. Press, p. 837.
24. According to Harold Green, who helped to draft the AEC charges, as reported in Stern (with the collaboration of H. P. Green), *The Oppenheimer Case.*
25. USAEC, *In the Matter of J. Robert Oppenheimer,* The M.I.T. Press, p. 710.

26. *Ibid.*, p. 726.
27. Stern, *The Oppenheimer Case*, p. 261. There he is quoted as saying, "the closest I ever came to being a Communist was voting for Franklin Roosevelt in 1932."
28. Smyth, *Atomic Energy for Military Purposes*.

Index

Index